Home Electrics

Question and Answers books are available on the following subjects:

Amateur Radio
Automobile Brakes & Braking
Automobile Cooling Systems
Automobile Electrical Systems
Automobile Engines
Automobile Fuel Systems
Automobile Lubrication
Automobile Steering & Suspension
Automobile Transmission Systems
BASIC Programming
Brickwork & Blockwork
Cameras
Car Body Care & Repair
Car Maintenance
Car Tools
Car Tuning
Carpentry & Joinery
CB Radio
Central Heating
Colour Television
Computers
Cycles & Cycling
Diesel Engines
Domestic Lighting
Electric Arc Welding
Electric Motors
Electric Wiring
Electricity
Electronics
Gas Shielded Arc Welding
Gas Welding & Cutting
Gems
GRP Boat Construction
Hi-Fi
Home Electrics
Home Insulation
Household Security
Integrated Circuits
Lathework
Light Commercial Vehicles
Microprocessors
Motor Cycle Care
Motor Cycle Electrics
Motor Cycles
Motoring Breakdowns
Painting & Decorating
Personal Computing
Pipework & Pipewelding
Plastering
Plumbing
Radio
Radio Repair
Refrigeration
Steel Boat Construction
Television
Transistors
Video
Videocassette Recorders
Wooden Boat Construction
Yacht & Boat Design

QUESTIONS & ANSWERS

Home Electrics

W. Turner, C. Eng., MIEE

Newnes Technical Books

Newnes Technical Books
is an imprint of the Butterworth Group
which has principal offices in
London, Boston, Durban, Singapore, Sydney, Toronto, Wellington

First published 1985

British Library Cataloguing in Publication Data

Turner, W.
Home electrics.—(Questions and answers)
1. Electric wiring, Interior
I. Title II. Series
621.319'24 TK3271

ISBN 0-408-01477-6

Photoset by Butterworths Litho Preparation Department
Printed in England by Whitstable Litho Ltd, Whitstable, Kent

Contents

Preface

This book will help the capable householder to understand the basic principles of home electrics and to undertake such jobs as replacing fuses or flexes, adding extra lights or sockets, or even complete rewiring. The information given should also be of interest to others, such as architects or students, involved with domestic electrical installations.

The savings to be made by doing such work yourself are substantial since the main cost of electrical work is for labour. The bulk of the work is not electrical and does not require special skills; jobs like lifting floorboards, chasing walls, running cables, fixing accessories etc. are well within the capability of any competent person.

However, electricity is potentially dangerous and must be treated with the utmost respect. The emphasis throughout the book is therefore on safety, which requires sound design in accordance with the relevant regulations, a high standard of workmanship and good-quality materials.

Don't take any risks; if in any doubt get assistance from a qualified person either in the form of advice or physical help. Before working on an electrical circuit, *always* switch off at the mains or remove the circuit fuse and, if working on an appliance, pull out the plug or remove the local fuse if there is one; if a long flex is involved, e.g. a lawnmower or a hedgetrimmer, take the plug with you so that it can't be re-inserted without your knowledge.

Provided the information in the book is carefully applied, the results should be both safe and satisfactory.

Advice is included on inspecting an existing installation to ascertain whether rewiring is necessary – this can be very useful when buying or selling property.

Grateful acknowledgement is made to the Institution of Electrical Engineers for permission to quote from the *Regulations for Electrical Installations*, and to the following manufacturers who have supplied illustrations and information on their products:

Ashley Accessories Ltd.; B & R Relays Ltd.; Crabtree Electrical Industries Ltd.; Findlay, Durham & Brodie Ltd.; Friedland Ltd.; Home Automation Ltd.; Midland Electric Manufacturing Co. Ltd. (Delta); MK Electric Ltd.; Santon Ltd.; Geo. H. Scholes Ltd. (Wylex); Square D Ltd.; Tenby Electrical Accessories Ltd.; Thorn Lighting Ltd.; Volex Electrical Products Ltd.

W. T.

1

Supply

What is electricity?

Electricity is a means of transferring energy (or power) from one point to another. It is not a *primary* source of energy such as coal, gas or oil, but is produced by burning such fuels and by certain other methods, including nuclear fission and wind or water power. At the points of use the electricity is converted into useful forms of energy such as light, heat or motive power.

How is electricity distributed?

The low-voltage system in the UK is known as '3-phase 4-wire', consisting of three phase conductors and a neutral. A voltage of

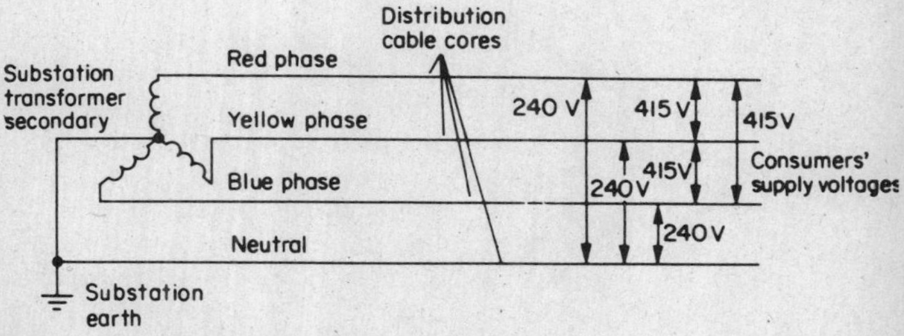

Fig. 1 *LV distribution system*

240 V exists between any phase conductor and the neutral whilst the voltage between any two phase conductors is 415 V (Fig. 1). The supply consists of alternating current (AC) at a frequency of 50 Hz (Hertz) or 50 cycles/second, which means that the current reverses in direction 50 times per second. The frequency in other countries may vary e.g. in the US it is 60 Hz.

A domestic dwelling is normally supplied at 240 V single phase, i.e. one phase conductor and the neutral. For large properties where high loads are expected (above, say, 24 kW), two or more phases may be supplied but all equipment on the premises is still operated at 240 V.

What is a 'live' conductor?

The term 'live conductor' includes any conductor which is connected to the phase or neutral side of the supply.

What is the intake position?

This is the position at which the incoming supply terminates in the building. At this position the electricity board install their main fuse, neutral link, meter and any other necessary items, e.g. timeswitches, when required. See Fig. 2.

The consumer's fuses, usually in the form of a consumer unit, (see Chapter 2) are usually fitted immediately adjacent to the meter and are connected by suitable 'tails', supplied by the consumer, to the electricity board's equipment.

The location of, and the arrangements at, the intake position should be agreed between the consumer and the electricity board.

It is now quite common for the electricity board's meters to be arranged so that they can be read from outside the premises.

It is not normally necessary for a domestic consumer to have a main fuse in addition to the fuse provided by the electricity board (see Fig. 3). The consumer unit includes a double pole main switch which controls all the outgoing ways. In some cases the main switch on the consumer unit is replaced by an earth-leakage circuit breaker (ELCB) (see Chapter 2) which performs the function of a main switch.

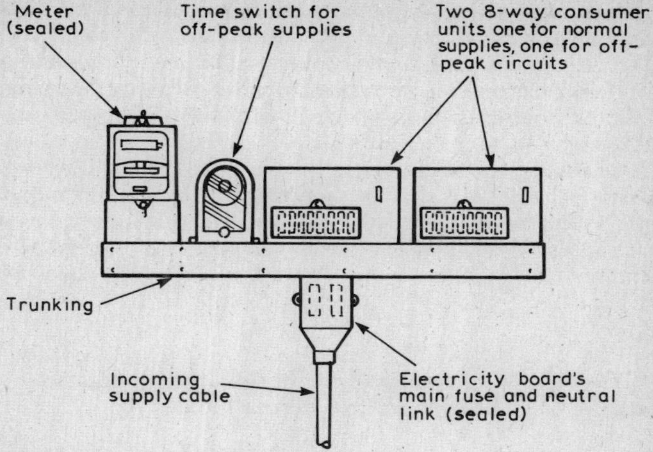

***Fig. 2** Typical consumer's terminal*

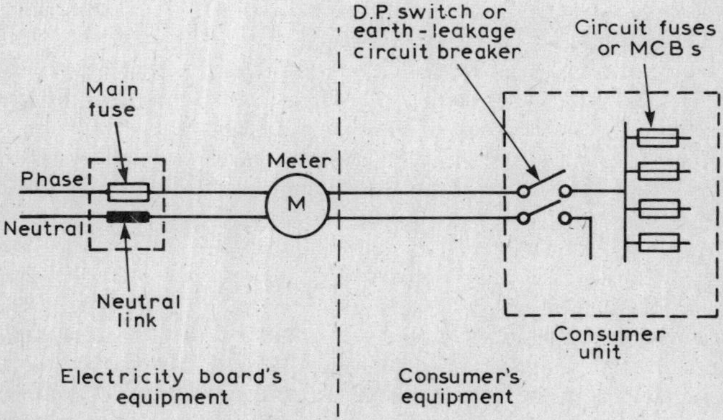

***Fig. 3** Sequence of control and protective equipment*

What regulations apply to domestic wiring?

In the UK, electrical installations should comply with the Regulations for Electrical Installations issued by the Institution of Electrical Engineers – usually called the IEE Wiring Regulations.

Before connecting the supply to an installation or an extension to an existing system the electricity board need to be satisfied that it complies with the Electricity Supply Regulations; if it does, they have no statutory power to refuse a supply. If the installation is in accordance with the IEE Wiring Regulations it automatically complies with the above Supply Regulations and there should be no difficulty. If the board decline to provide a supply, they are obliged to state the reason in writing.

In the UK there is no restriction on persons carrying out electrical work as there is in some countries, e.g. Australia, where it may only be done by registered electricians.

What voltage limits are applicable?

Supply voltage. The Electricity Supply Regulations for the UK require that the voltage at a consumer's intake position must not vary by more than 6 per cent above or below the nominal voltage. Thus on a nominal 240 V supply the voltage should be within the range 225–255 V.

Consumer's premises. The IEE Wiring Regulations require that on a consumer's premises the voltage drop from the intake position to any point on the installation under full load conditions must not exceed 2.5 per cent of the nominal voltage. Thus on a 240 V supply the maximum permissible volt drop is 6 V.

How is the consumption of electricity measured?

The meter provided by the electricity board records the consumption in units.

A unit is a kilowatt-hour (kWh) which means a load of 1kW for one hour or its equivalent, e.g. 0.5 kW for two hours or 2 kW for half an hour, etc. A kilowatt is 1000 watts.

Typical loadings of commonly-used domestic appliances are given in Table 1.

Table 1 Power consumption of various appliances

Appliance	*Loading in watts*
Lamps (filament)	15, 25, 40, 60, 75, 100, 150
Lamps (fluorescent)	15, 20, 30, 40, 50, 65, 75, 85
Television sets	
black-and-white	75
colour	150
Fires and heaters	500–3400
Water-heaters, immersion heaters	750–3000
Wash boilers	3000
Washing machines	3000
Spin-driers	300–500
Drying cabinets	2000–3000
Refrigerators	300–400
Cookers	6000–8000
Portable appliances (e.g. irons, vacuum cleaners, etc.)	10–3000

What tariffs are available?

There are normally two tariffs available to domestic consumers although these may vary between different areas: (a) standard domestic tariff and (b) off-peak (Economy 7).

The former applies when the off-peak tariff is not used and it involves a fixed charge per quarter plus a further charge for each unit used.

On the Economy 7 tariff there is again a fixed quarterly charge but *all* units used during a 7-hour period during the night are charged at a low rate whilst units used during other periods are charged at a higher rate.

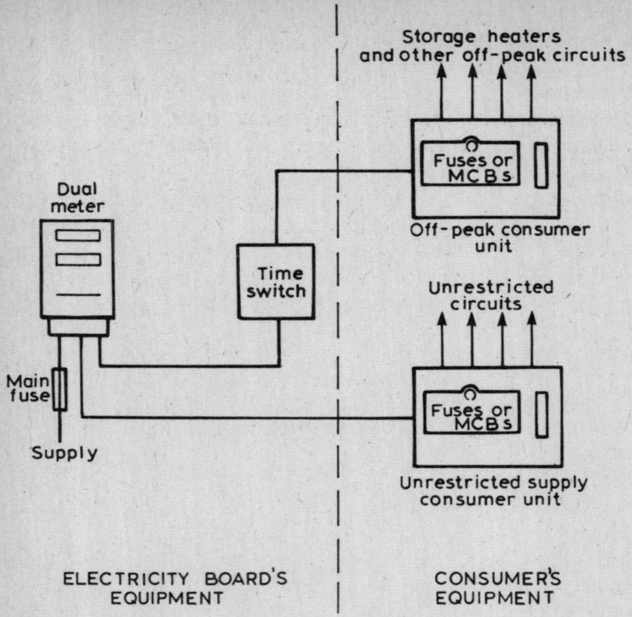

Fig. 4 *Typical connections for off-peak supply*

The meter used on the Economy 7 tariff has two sets of numerals; a timeswitch, supplied by the electricity board, switches the appropriate set into use according to the time of day. To facilitate the connection of consumer's equipment, the electricity board provide two sets of terminals to the consumer; one set is permanently 'live' but the other set is 'live' only during the 7-hour off-peak period. Appliances which are needed *only* during the off-peak period, e.g. storage heaters, are connected to the off-peak terminals via a separate consumer unit (described in Chapter 2). Fig. 4 shows typical connections with an off-peak supply.

What is earthing?

As shown in Fig. 1, the neutral conductor of the supply is connected to earth at the transformer which feeds the low-voltage (LV) system. There is thus a voltage of 240 V between each phase conductor and earth.

If a phase conductor is in contact with any metalwork on the installation which is not earthed (e.g. the framework of an electric fire or other appliance) this metalwork will be at a voltage of 240 V with respect to earth. If the metalwork is then touched by a person who is in contact with earth, a severe, and possibly fatal, shock will result. See Fig. 5.

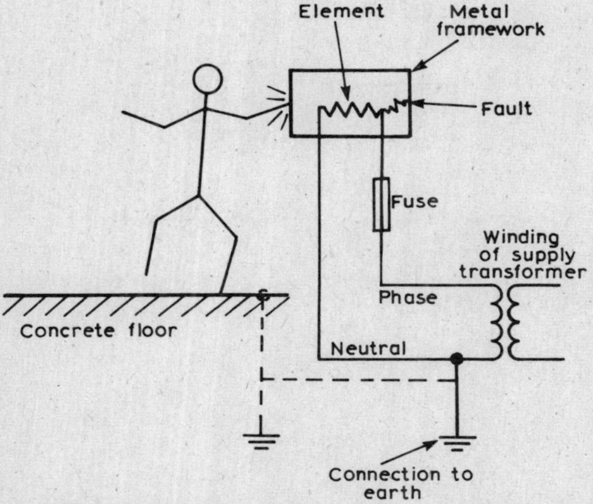

***Fig. 5** How a person may receive a shock from a faulty appliance*

However, if the metalwork is connected to earth the above situation cannot arise since there will be no voltage with respect to earth, and also the excess-current protective device for the circuit should operate and disconnect the supply. See Chapter 2.

The metalwork of an appliance such as a fire can make contact with a live conductor due to a damaged flex or other faulty condition. This situation arises if the earth wire in a plug becomes detached from its terminal and touches the phase terminal; this removes the earth from the metalwork and when the plug is inserted into a socket the metal becomes live since it is connected

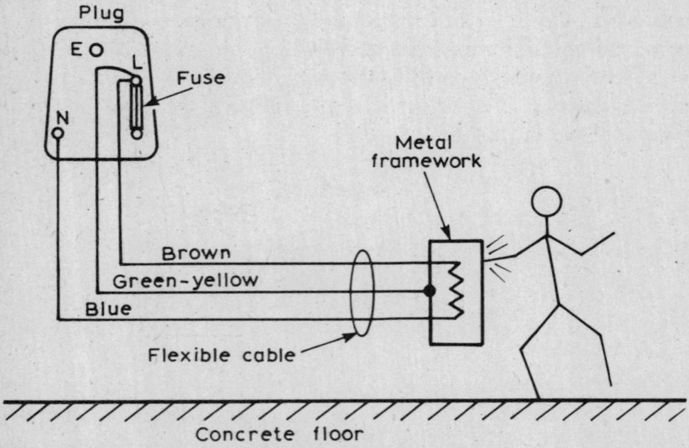

Fig. 6 Showing risk of shock if earth wire becomes detached from plug terminal and touches phase terminal

to the earth wire in the plug. In the above case the fault is not apparent since the appliance functions normally, but the excess-current protection for the circuit does not operate since the phase wire is not connected to earth (Fig. 6). Protection against this situation can only be provided by an earth-leakage circuit breaker – see Chapter 2.

How is earthing achieved?

It is the responsibility of the consumer, not the electricity board, to provide an earth connection on consumers' premises. However, when the supply is underground, which applies in the majority of cases, the electricity board usually provide an earth terminal at the intake position. This is called the main earthing terminal and is connected to the metal sheath and/or armour of the supply cable. The sheath and armour provide contact with the general mass of earth and are also connected to the earthing system at the transformer and to the neutral point of the transformer (see Fig. 1). This usually provides a satisfactory earth connection, but if not, alternative arrangements must be made by the consumer as indicated below.

The earth terminal on the consumer's premises is connected to the metalwork of all electrical appliances on the premises by

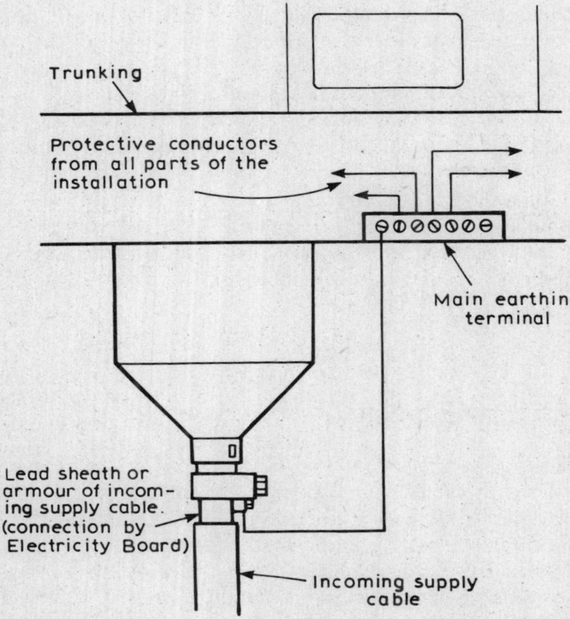

Fig. 7 *Earthing connections at consumer's intake position*

means of the protective conductor (see page 11); this is connected to the earth terminals of all socket outlets and the metalwork of all fixed appliances. See Fig. 7.

It is *not* permissible to use metallic water- or gas-pipes as the *sole* means of earthing.

If no earth connection is provided by the board, the consumer must provide an earth electrode system. This normally consists of buried metal rods, strip or plates to provide contact with the general mass of earth. However, it is not normally practicable to obtain a permanent satisfactory earth connection by this means, and the IEE Wiring Regulations require that when the sole means of earthing is by a local earth electrode system, i.e. there is no metallic earth connection back to the supply transformer, the socket outlets must be protected by a residual-current-operated earth-leakage circuit breaker (ELCB) operating at not more than 30 mA. ELCBs are described in Chapter 2. However, in many cases where the mains supply is overhead, the electricity board adopt the protective multiple earthing (PME) system (described later in this chapter), and the foregoing requirement does not then apply.

An ELCB, as described above, is also required to protect socket outlets which may be used to supply portable equipment outdoors via a flexible cable, and such sockets must be labelled 'For equipment outdoors' (IEE Regulations 471-12 and 514-8).

What labelling is required for earth connections?

The IEE Wiring Regulations (514-517) require that a permanent label, durably marked 'Safety electrical connection – do not remove' in letters at least 4.75 mm high, should be fitted (a) at the point where the earthing conductor is connected to the earth electrode system (when provided) and (b) at all points where a bonding conductor is connected to metallic pipes etc. A similar label should preferably be provided on the cable from the consumer unit to the main earthing terminal.

What is the protective conductor?

This is the means by which the parts of an installation which require to be earthed are connected to the consumer's main earthing terminal. It may consist of a metal conduit system; a metal cable sheath or armour; a separate conductor in a PVC-insulated, PVC-sheathed cable; or a separate conductor not enclosed with others.

The protective conductor must be taken to *all* outlet points on the installation including socket outlets, lighting switches, lighting outlets etc; this applies even when the protective conductor cannot be connected initially at the outlet point, e.g. when all-insulated fittings are used.

Domestic wiring usually consists of flat PVC-insulated, PVC-sheathed cables which include a bare earth conductor (see Chapter 3). At all termination points the length of bare conductor within the outlet box, accessory or consumer unit must be enclosed in insulating sleeving, coloured green-yellow.

What is protective multiple earthing (PME)?

On this system the main earthing terminal at the intake point is connected by the electricity board to the neutral conductor (Fig. 8). This means that a phase-to-earth fault on the premises becomes a phase-to-neutral fault.

In the UK, the wiring on the premises to all outlet points is normal, i.e. the protective conductors and the neutral are separate from each other for all wiring on the consumer's side of the consumer unit, and the neutral must not be earthed at any other point.

PME was originally used only in rural areas having overhead supplies where it was difficult or impossible to obtain satisfactory earthing. However, it is now being widely applied even in urban areas where the supply is underground.

Bonding of metalwork as described below is particularly important with PME working and details of this should be agreed with the electricity board.

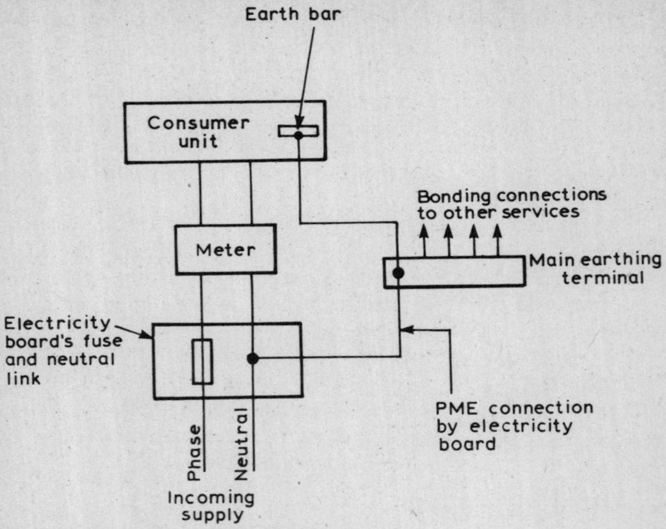

Fig. 8 Connections for PME system

The PME system provides a low-impedance earth path back to the supply transformer so that the fuse or MCB should operate quickly when a phase-to-earth fault occurs.

PME working can only be adopted at the instigation of the electricity board and the decision on its use cannot be made by the consumer.

What is bonding?

Bonding refers to the connection to each other, and to the main earthing terminal, of metallic pipes entering the building and other exposed metalwork which could become live and could be touched. This includes gas and water service pipes, metal sinks and baths, radiators, and structural steelwork and metal ducting.

The bonding connections to the incoming gas or water service pipes are termed *main* bonding connections and should be as near as possible to the points of entry into the building but on the consumer's side of any insulating sections, meters or stop cocks (Fig. 9). This may mean a long run of cable if the other services do

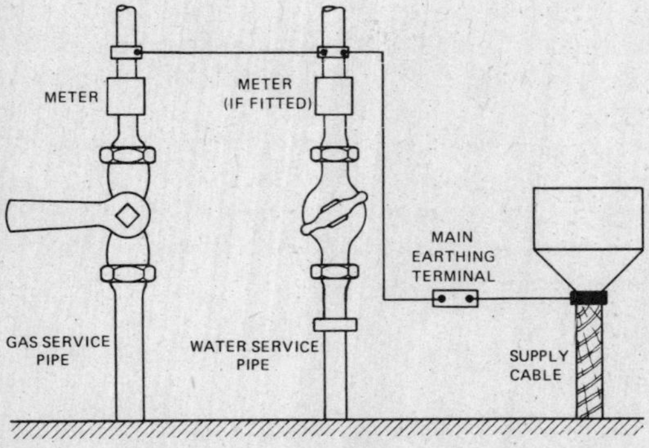

Fig. 9 *Bonding of services. Connection to gas service pipe should be within 600 mm of gas meter*

not enter the building at the same point as the electrical supply. The cable for this bonding should be PVC-insulated, single-core, coloured green-yellow; the minimum size should be 6 mm^2 if the supply is not PME (see above) or 10 mm^2 if the PME system is in use. However, it is recommended that 10 mm^2, minimum, should be used in all cases so as to provide for future PME working. If in doubt consult the electricity board.

The object of the above bonding is *not* to provide an earth connection to the premises but to ensure that the pipes and metalwork cannot become live.

In addition to the above, *supplementary* bonding may also be required to exposed conductive parts such as metal baths or sinks, towel rails, metal pipes, radiators etc. In most domestic situations, heating pipes, radiators, hot water cylinders, hot and cold water pipes etc. are automatically bonded to the main earthing terminal via the metal piping and metal-to-metal joints but, if this is not the case, bonding should be applied.

However, metal baths or sinks may not be electrically connected to the pipework due to the use of compound around the taps or for other reasons and, to make sure, these items should be bonded by separate conductors, which should be connected to the protective conductor (see above) at a suitable point or to the main earthing terminal. Some metal baths include an earth terminal. Towel rails in bathrooms or showers should be treated similarly.

The above requirements are particularly important in rooms containing a fixed bath or shower, and in kitchens.

The minimum cable size for this supplementary bonding should be 4 mm^2. Other special requirements applicable to bathrooms are covered in Chapter 2.

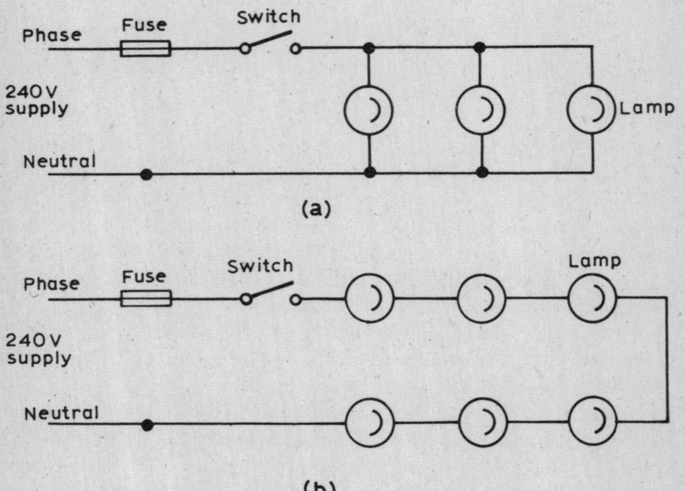

Fig. 10 *Connection of lamps: (a) in parallel, (b) in series*

How are current-using appliances connected to the supply?

Appliances are almost invariably connected in *parallel* i.e. a phase and neutral connection of negligible impedance is taken to each appliance. The main exception to this is that 'Christmas tree' lights are often connected in *series*. Fig. 10 shows the two systems.

With the series system, if one lamp fails all the lamps are extinguished; also the number of lamps per circuit depends on the voltage of the lamps and the supply voltage, e.g. on a 240 V supply using 12 V lamps there must be 20 lamps per circuit.

With the parallel arrangement all lamps and other appliances must be suitable for mains voltage; also the failure of one lamp or other item will not affect the others.

2
Protection and safety

How is protection provided against excess current?

If the current flowing in a circuit is greater than the rated capacity of the circuit the cables or equipment may be damaged and there may be a risk of fire.

Protection against this situation is provided by fuses or miniature circuit breakers (MCBs) which disconnect the supply from the circuit when the current reaches a pre-determined value.

The rating of the fuse or MCB should normally not exceed that of the circuit which it supplies, e.g. if the cables and equipment have a maximum rating of 30 A, the rating of the fuse or MCB should not exceed this value.

What is a fuse?

A device which disconnects the circuit when an excessive current flows. In consumer units (see page 26) or fuseboards the fuse normally consists of (a) a fixed base to which the fixed wiring is connected and (b) a removable link which houses the fuse element and which plugs into the fixed base. Fuses for 13 A plugs clip directly into spring contacts in the plug.

The *current rating* of a fuse is the maximum current it will carry continuously without undue temperature rise or deterioration.

The *fusing current* is the minimum current required to blow the fuse.

The ratio of fusing current to current rating is called the *fusing factor*, e.g. if the values are 40 A and 20 A respectively the fusing factor is 2.

In domestic installations the fuses are normally fitted in consumer units and the sizes commonly used are 5 A, 15 A, 20 A, 30 A and 45 A.

What is a rewirable fuse?

In the rewirable type the fuse wire usually passes through the body of the carrier to terminals at each end – see Fig. 11. The

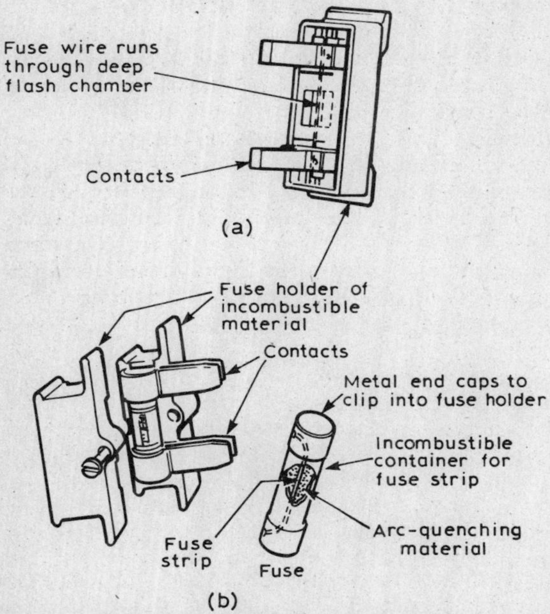

***Fig. 11** Fuses: (a) rewirable type, (b) cartridge type*

fusing factor for this type is usually 2, i.e. a 30 A fuse will blow at approximately 60 A.

Rewirable fuses have the advantage of being cheap, and replacement is simple, costing virtually nothing. Disadvantages are that it is very easy to insert the wrong size of fuse wire, which may lead to danger; also the fuse wire can deteriorate due to oxidization etc.; there is also a risk of flashover at high fault currents.

What is a cartridge fuse?

In the cartridge fuse, the fuse element is totally enclosed in a sealed ceramic tube filled with a fireproof material such as quartz and is connected to fixed contacts at each end of the tube. See Fig 11. The fusing factor of a cartridge fuse for domestic use is normally 1.5, i.e. a 30 A fuse will blow at about 45 A.

Advantages of cartridge fuses are that their operating characteristics are more precise, they operate more quickly and require less current to blow than the rewirable type. Since the fuse wire is totally enclosed, there is no risk of flashover and they do not deteriorate in use. In general, the different fuse ratings have different dimensions so there is less risk of the wrong fuse rating being inserted.

Disadvantages are that they are more expensive than the rewirable type and, since they cannot be rewired, a new fuse must be used each time a fuse blows; thus spare fuses must be held for each rating in use.

Cartridge fuses (of a different size to those used as circuit fuses) are used widely in 13 A fused plugs and fused connection units on ring or radial circuits feeding socket outlets to BS 1363. See Chapter 5.

How can I test a fuse?

Both the rewirable and cartridge types can be tested by means of either a bell and battery set (see Chapter 9, page 126) or a metal-cased battery-operated torch.

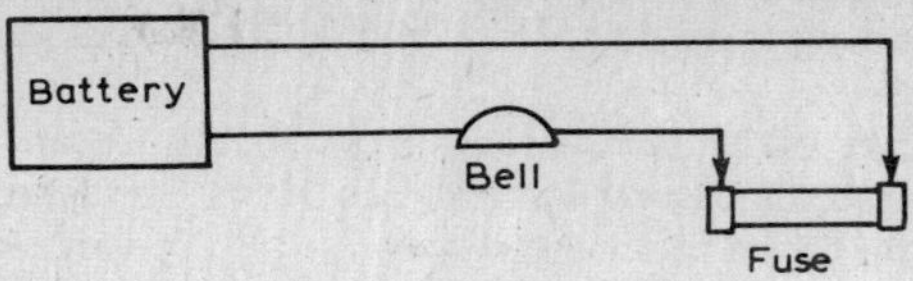

(a)

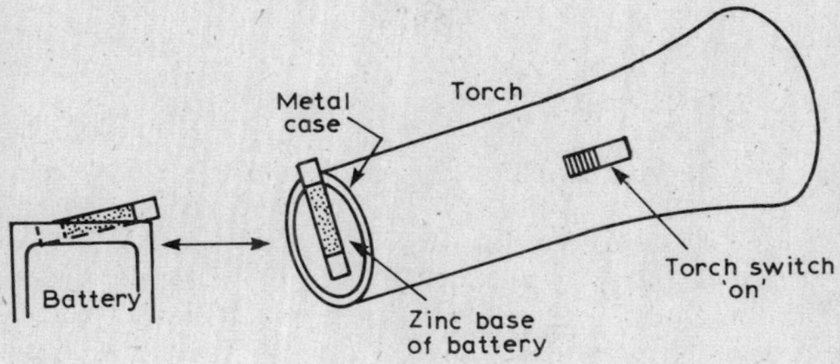

(b)

Fig. 12 Testing fuse: (a) using bell and battery set, (b) using metal-cased torch

With a bell and battery set, tap the two test leads on to the fuse terminals and if the bell rings the fuse is in order; if not, the fuse is faulty.

For testing with a torch, place the torch switch to 'on', remove the base and hold the fuse so that it makes contact simultaneously with the base of the battery and the metal case, taking care that the battery is pressed towards the top of the torch. If the fuse is sound, the torch will light.

What is a miniature circuit breaker?

A miniature circuit breaker (MCB) is an electro-mechanical device (usually single pole) which operates and disconnects the circuit when the current reaches a pre-determined value. MCBs may thus be used in lieu of fuses and can be fitted in consumer

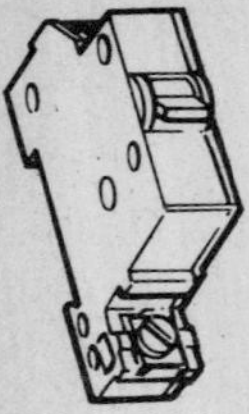

(a)

***Fig. 13** Typical units: (a) miniature circuit breaker, (b) standard consumer unit, (c) split-load consumer unit (Wylex Ltd.)*

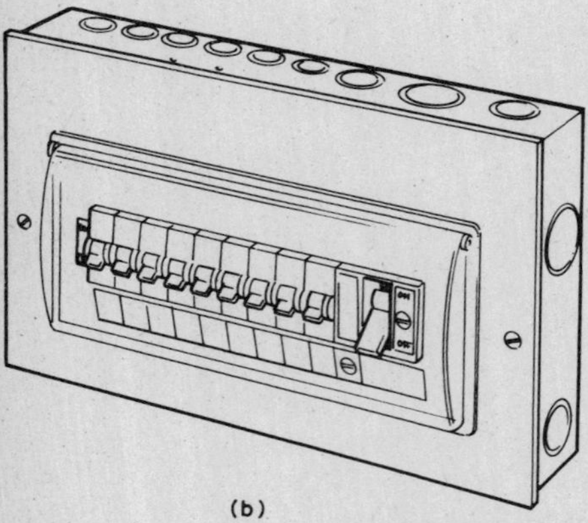

(b)

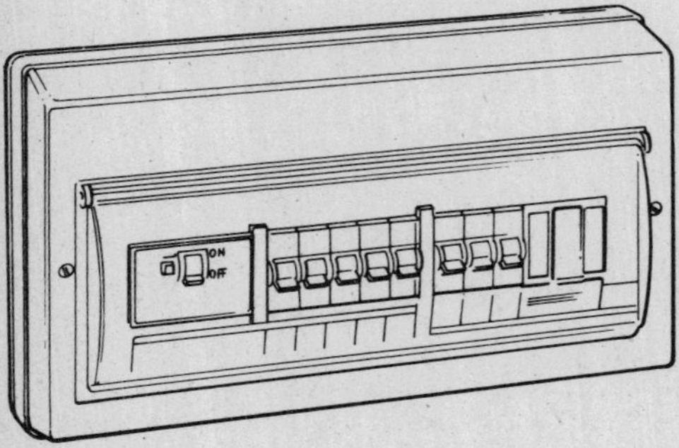

(c)

units in a similar manner – see page 26. Fig. 13 shows a typical MCB and consumer unit.

The ratings used on domestic work are generally as for fuses as stated earlier.

An MCB will normally operate at 1.25 times its rated current, e.g. a 20 A MCB operates at 25 A compared with 30 A for a cartridge fuse or 40 A for a rewirable type of the same rating.

Other advantages of MCBs are that they can be quickly reset by hand after operation without any cost and they will not reclose if the fault still exists. Also an MCB which has tripped can be readily identified since the dolly moves to the 'off' position. MCBs permit individual circuits to be switched off easily if a dwelling is left unoccupied, whilst leaving essential circuits switched on, e.g. freezers etc.

MCBs are slightly more expensive than fuses but this is offset by the fact that no cost is involved in resetting them after operation; and this is a very quick and simple process. They are now being used to an increasing extent in lieu of fuses.

How is protection provided against earth leakage currents?

The current in an electric circuit should normally flow only in the phase and neutral conductors and the current-carrying parts of appliances. Leakage currents to earth occur if the above items come into contact with an earthed object.

Earth leakage currents are undesirable since they may cause a dangerous voltage on exposed metalwork. They are also a fire risk since the leakage current may take a path not designed for the purpose, causing overheating.

Protection against earth leakage currents is provided by (a) the excess-current protection described earlier or (b) earth-leakage circuit breakers (ELCBs).

It is obvious that when the protection is by means of fuses or MCBs, the protective device will operate only if the earth leakage current reaches the value required to operate the device; this means that the impedance of the fault path must be sufficiently low to permit the required current to flow. If this condition is not met this method of protection will obviously be ineffective and the leakage will persist.

In such situations the protection against earth leakage can be provided by using ELCBs; these can be arranged to operate very quickly with extremely low leakage currents, far less than the values necessary to operate a fuse or MCB. Thus the protection against electric shock provided by an ELCB is considerably greater than that by a fuse or MCB, e.g. the operating values for a 30 A rewirable fuse, cartridge fuse and MCB are 60, 45 and 37.5 A respectively whereas an ELCB will operate at 30 mA.

As indicated in Chapter 1, ELCBs are compulsory in certain situations but they are now being much more widely used even when not mandatory. See page 10.

What is an ELCB?

An ELCB detects earth leakage current and disconnects the supply when the current reaches a pre-determined value. It consists essentially of an operating coil and a trip mechanism which operates contacts controlling the supply to the circuit concerned.

There are two basic types: (a) fault-voltage operated and (b) residual-current operated. On the former type the operating coil is connected in series with the consumer's main earthing lead and operates when earth leakage occurs. However, this type has certain drawbacks and has now been almost entirely superseded by the residual-current type for domestic work; it is not therefore described further in this book. The IEE Wiring Regulations express a preference for the residual-current type – the modern name for this is 'residual current device' (RCD).

How does an RCD work?

The phase and neutral currents of the circuit being protected are fed through separate windings of a transformer as shown in Fig. 14 and another winding controls the tripping mechanism. The contacts are normally held closed by a spring mechanism and open when this is tripped.

Under normal conditions the phase and neutral currents are equal and the transformer connections are such that they neutralize each other. If earth leakage occurs from either the phase or neutral side of the circuit, e.g. by a person touching a live conductor, the currents are unbalanced by an amount equal to the leakage current and if this reaches the rated tripping current of the RCD, the tripping mechanism operates and disconnects the circuit.

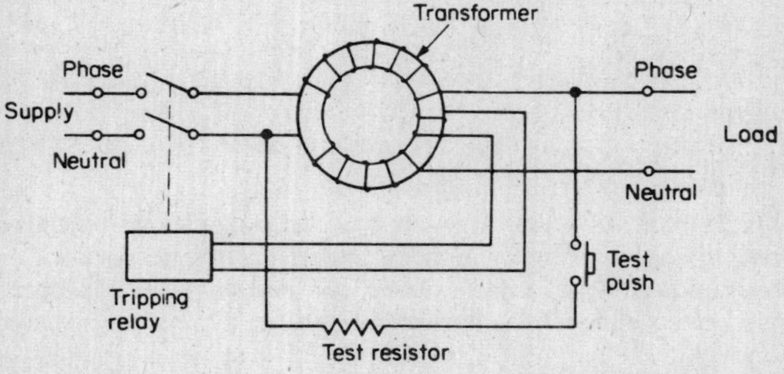

***Fig. 14** Connections for residual-current operated earth-leakage circuit breaker (residual current device)*

A test button is included on the RCD as shown in Fig. 14 which simulates a fault and enables a quick test to be made at any time to ensure that the trip mechanism is working satisfactorily. This test should be made periodically, say at monthly intervals.

An RCD will operate if a live terminal or conductor is touched by an earthed object or person even if there is no contact with the earthed metalwork, provided the leakage current exceeds the operating current of the RCD. However, the earthing arrangements, protective conductors and excess-current protection must still be provided in the normal manner.

The RCD must be reset manually after operation and it cannot be reset if the leakage still exists.

How are RCDs used?

RCDs (Fig. 15) are available in various forms:

(a) As a separate unit for wiring into the circuit as required.
(b) Combined with a single or twin socket outlet for connection to fixed wiring.
(c) Combined with a socket outlet, with flexible cable for connection via a 13 A plug to any standard socket.
(d) Combined with a 13 A plug for permanent connection to a flexible lead on a portable appliance in lieu of a normal 13 A plug.
(e) Combined with a 13 A plug and socket outlet. The unit is plugged into a standard socket outlet and the appliance to be protected is plugged into the RCD unit.
(f) Incorporated in a consumer unit and controlling some or all of the outgoing ways. See page 26.

The type of RCD used depends on individual circumstances and preference. As a minimum, earth leakage protection by means of a RCD operating at not more than 30 mA *must* be provided (1) when the sole means of earthing is by local earth electrodes and (2) for socket outlets, which are intended to supply portable appliances outdoors via flexible cables (see Chapter 1).

At the other extreme a common RCD may be provided at the intake position to protect the whole installation; this means, however, that a leakage on any circuit or appliance will disconnect the whole dwelling, including lighting. Also it may not be advisable to connect circuits supplying such items as freezers, greenhouse heating, fish tanks and possibly central heating controls to an RCD which also feeds other items.

It can be argued that the risk of leakage is greater on socket outlet circuits than on permanently-wired equipment; thus a compromise may be to provide earth leakage protection only on the socket outlet circuits and the compulsory items mentioned above. Alternatively, the protection can be provided only in especially dangerous situations such as garages, outbuildings etc.

If type (d) above is used this ensures that earth leakage protection is *always* provided when the particular appliance is

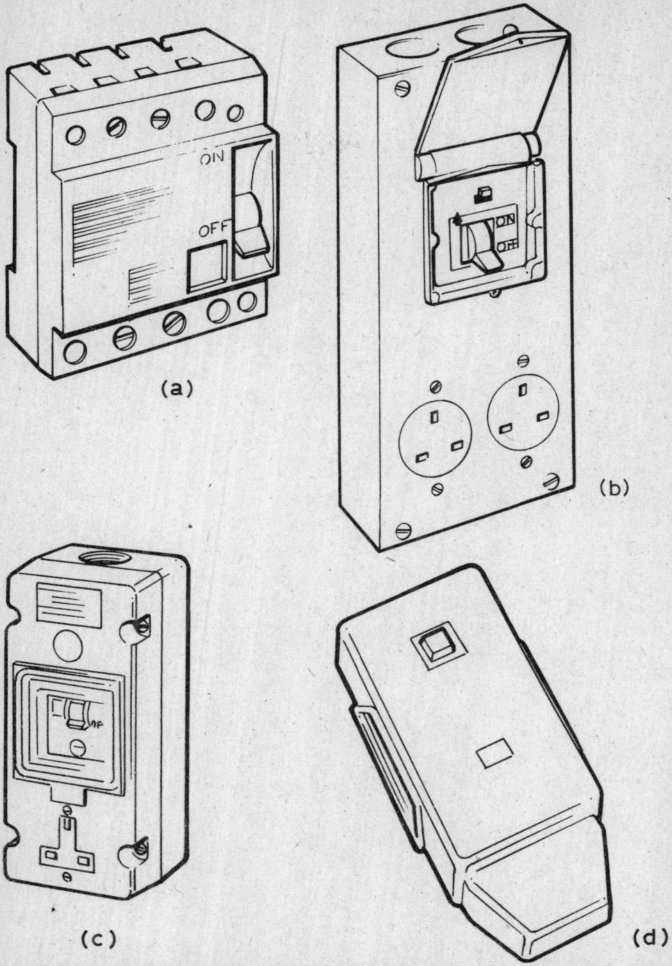

Fig. 15 *Typical residual current devices: (a) separate unit (MK Electric Ltd.), (b) and (c) combined with socket outlets (Wylex Ltd.), (d) combined with 13A plug (B & R Relays Ltd.)*

used but it means that the RCD can only serve one appliance. However, if the RCD is connected by a short length of flex to a standard socket, it can be used with any appliance.

RCDs do not give protection against excess current or short circuits between phase and neutral and this protection must be provided in the normal manner; the requirements for earthing must also still be met as described in Chapter 1.

What is a consumer unit?

A consumer unit consists of a double-pole (DP) main switch, usually 60, 80 or 100 A rating, and the required number of outgoing circuits, often called 'ways'. Fig. 13 shows typical units.

On some units the main switch is replaced by an RCD which provides earth leakage protection on outgoing circuits and also acts as the main switch.

Split-load consumer units are also available on which some ways are fed via a common RCD and the others are fed direct from the main switch; the main switch controls *all* the outgoing circuits. Fig. 13 shows a typical unit of this type with five ways via the RCD and three direct from the main switch. This type of unit permits some circuits, e.g. socket outlet ring circuits, to have earth leakage protection whilst others can be either fed direct or via separate RCDs.

Consumer units are also available which, in effect, combine two consumer units in a common case. Each bank of circuits is controlled by a separate main switch, fed by a separate supply. These units are intended for use with off-peak, Economy 7, tariffs which usually involve two consumer units.

All consumer units include neutral and earth bars and a main earth terminal and they can be flush or surface-mounting with metal or insulated enclosures. A metal enclosure should be used for a conduit system whilst for a PVC-sheathed cable system the insulated type is usual but a metalclad unit may be used if desired, suitable grommets being fitted where the cables enter the enclosure.

How are consumer units installed and wired?

Consumer units are usually installed at the intake position adjacent to the electricity board's meter. Surface-mounted units should preferably be fitted on a separate wood or fibre board at least 13 mm thick mounted on suitable spacers so as to give about 25 mm space behind the board. This allows the wiring to be concealed and prevents rusting of metal enclosures. Cables can then enter the consumer unit at the top or bottom or from behind. The board should be generously sized so as to allow space for possible future items e.g. timeswitches or additional consumer units.

A cooker of more than 3 kW nominal loading should be on a separate circuit, the loading being calculated in accordance with the details in Chapter 9. A large cooker may require a 45 A circuit. Shower units and other water heaters of 3 kW or more should also be on separate circuits.

Each ring or radial circuit feeding socket outlets should be fed by a separate fuse or MCB rated at 20 or 30 A – see Chapter 5. For socket outlet ring circuits the two phase conductors should terminate on the same fuse or MCB. When there are two or more ring circuits, care must be taken that the wires are not mixed.

Domestic lighting circuits are normally fed via 5A fuses or MCBs and, except in very small properties, two circuits should normally be provided so that the dwelling will not be in darkness should one circuit fail.

In terminating the cables in the consumer unit the neutrals and protective conductors should be in the same order as the phase connections. The sheathing of PVC-sheathed cables must protrude inside the enclosure so that there is no unsheathed wiring outside. All bare protective conductors within the unit should be enclosed in green-yellow PVC sleeving.

Suitable permanent labelling should be provided, indicating the circuits fed by the individual ways, and all fuses and MCBs should be correctly colour-coded.

The shield over the main incoming terminals must be replaced immediately the main supply cables have been terminated, since these terminals are live even when the switch is off.

Any unused knockouts in the enclosure should be left intact or closed by rubber plugs to prevent the entry of insects or dirt.

How is the mains supply connected?

It is the responsibility of the consumer to provide the short connections (or 'tails') between the consumer unit and the meter although the actual connection at the meter is made by the electricity board. The cables should be single-core PVC-insulated and PVC-sheathed. The main earth connection from the earth bar in the consumer unit to the main earthing terminal should also be provided by the consumer using single-core PVC-insulated cable coloured green-yellow. All these cable sizes should be agreed with the electricity board.

Why must the neutral conductor be unfused?

If a fuse were connected in the neutral it would still be necessary to have a fuse in the phase conductor and the two fuses would then be in series. If a short circuit or overload occurred it is quite possible that the neutral fuse would blow whilst the fuse in the phase conductor remained intact. This condition would be dangerous since any appliance connected to the circuit, e.g. a vacuum cleaner or television set, would be live even though this would not be apparent.

A similar situation could arise if the neutral fuse were removed and the fuse in the phase conductor left in circuit.

The above remarks also apply when MCBs are used in lieu of fuses.

All consumer units, fuseboards, socket outlets etc. are connected during manufacture to comply with the above and, provided the wiring is connected correctly, the fusing will be correct.

Why must single-pole (SP) switches not be connected in the neutral?

The reasons in this case are similar to those applicable to fuses as stated above. If a SP lighting switch is connected in the neutral conductor the phase conductor will obviously be connected directly to the lighting outlet. Similarly, if the connections to a switched socket outlet are reversed, the phase conductor will be permanently connected to an appliance plugged into the socket even though the switch on the socket is off. Both these situations are dangerous.

A switch in the neutral is only permissible if it is linked to a switch in the phase conductor, i.e. by a double-pole (DP) switch.

What special precautions must be taken in bathrooms?

The IEE Wiring Regulations include the following requirements for rooms containing a fixed bath or shower:

1. There must be no socket outlets and no provision for connecting portable appliances other than electric shavers – see below. If a shower is fitted in a room other than a bathroom this requirement applies in an area within 2.5 m of the shower; this point should be watched when a shower is fitted in a bedroom which may contain existing socket outlets.
2. Shaver supply units complying with BS 3052 or 3535 (see Chapter 5) may be installed and the earth terminal must be connected to the protective conductor. (Shaver socket outlets which do not comply with the above are available which may be used in areas other than bathrooms – see Chapter 5).
3. All parts of a lampholder within 2.5 m of the bath or shower must be constructed of, or shrouded in, insulating material, and bayonet-type lampholders must be fitted with protective shields to BS 5042. Alternatively, totally enclosed lighting fittings may be used.
4. Lighting fittings with flexible cord pendants should not be used and there must be no open lampholders into which a

lampholder adaptor could be plugged to connect a portable appliance such as a hair dryer or electric shaver.

5. A lighting fitting should not be fitted over the bath where it could be touched by a person standing in the bath.
6. All exposed metalwork must be bonded together and to the protective conductor – see Chapter 1.
7. Every switch or other means of control or adjustment must be situated so as to be normally inaccessible to a person using a fixed bath or shower. This does not apply to electric shaver units complying with 2 above, to the insulating cords of cord-operated switches, nor to controls of *instantaneous* water heaters which comply with BS 3456, Section 3.9. It does, however, apply to the controls of normal immersion heaters, including the selector switch on a 'dual' heater – see Chapter 5.
8. No stationary appliance having heating elements which can be touched may be installed within reach of a person using the bath or shower; the sheath of a silica-glass-sheathed element is regarded as part of the element.

What precautions should be taken when installing wiring and equipment?

- Make sure that all work is in accordance with the latest issue of the IEE Wiring Regulations.
- Don't attempt work unless you are sure it is within your capability. If in doubt, get expert advice.
- Before starting any work on fixed wiring or equipment, turn off the main switch on the consumer unit and remove the circuit fuse. You may then need to turn on the main switch again so as to obtain lighting or power but, if not, leave it off. Also turn off any local switches on the circuit concerned and check that the circuit is, in fact, dead before working on it. Keep the fuse in your pocket so as to prevent anyone else replacing it.
- Before working on portable equipment *always* take out the plug from the socket outlet. In the case of fixed equipment fed

via a fused connection unit, *always* remove the fuse from the connection unit.

- Be particularly careful when doing work which involves switching the supply on and off for testing or checking. Check at each stage that the supply is *off* before touching wires or terminals which could be live.
- Before making fixings in plastered walls for such items as pictures, shelves etc., try to deduce the location of any concealed cables so as to avoid damage. They *should* be run vertically or horizontally to switches or socket outlets etc., but this cannot be guaranteed. If in doubt, turn off the main switch on the consumer unit until the work is completed.
- Use good-quality materials and a high standard of workmanship.

What precautions should be taken with flexes?

- Avoid long flexes and don't run flexes under carpets. An adequate number of socket outlets will help in this.
- Avoid joints in flexes as far as possible, but if you can't, use purpose-made connectors, *not* insulation tape.
- Ensure that the current ratings of all flexes are at least equal to the ratings of the fuses or MCBs through which they are fed.
- Don't use flexes for fixed wiring.
- Preferably, use sheathed flex rather than the twisted type, particularly for portable appliances.
- Don't use a long flex on a power tool; the plug and socket outlet should be within reach (about 2 m) when the tool is being used. If the socket outlet is not within reach, use an extension lead terminating on a socket outlet.
- After using a portable appliance outdoors, such as a lawnmower or hedgecutter, examine the flex carefully for any damage; flexes in these situations are subject to rough usage.
- Use heat-resisting flexible cable for immersion heaters and other heating appliances and for pendant-type luminaires exceeding 60 W.
- Check periodically that the fixing screws for all plug covers are tight; if the cover comes off, live terminals are exposed.

What precautions should be taken when using electrical equipment?

- Don't use plug adaptors unless unavoidable. If you have to use them, make sure they are fused. On an adaptor for use with 13 A plugs the *input* should be fused at not more than 13 A whilst on those intended for 15 A round-pin plugs the output circuits should be fused at 5 A maximum.
- Never fill an electric kettle at the sink whilst it is still plugged into a socket outlet and don't leave a kettle flex plugged into a socket with the other end free.
- At night or when leaving the house unoccupied, isolate appliances such as television sets etc. by switching off at the socket outlet and, preferably, also pulling out the plug; don't rely solely on the switch on the appliance.
- Never leave a radiant fire in a position where it could cause a fire or damage if switched on, e.g. facing furniture, curtains, woodwork or a wall. Remember the plug may be inserted into a socket outlet or the socket switched on in error, or by a child.
- Don't have a radiant fire arranged to be switched on automatically by a timeswitch.
- Don't have staircase or landing lights controlled by a delay switch (see Chapter 4) which suddenly switches off the lights. Use a switch which dims the lights gradually.
- It is preferable that the lighting should not be controlled by an ELCB (see Chapter 2) which also feeds other circuits, since the operation of the ELCB would suddenly extinguish all the lights.
- Don't let curtains or other items come into contact with tubular heaters.
- Don't put articles to dry on convector heaters or storage heaters.
- Never feed any appliance or equipment via a cable (flexible or other type) whilst the cable is coiled on a drum; otherwise the cable may overheat.

3
Wiring systems

What are the main systems for domestic wiring?

The three systems normally used are:

(a) PVC-insulated, PVC-sheathed cables with earth conductor.
(b) PVC-insulated, non-sheathed, single-core cables enclosed in metallic or non-metallic conduits.
(c) Mineral-insulated copper-sheathed (MICS) cable.

Systems (b) and (c) require special tools and skills for installation and are more expensive than system (a); they are not normally used for domestic work.

Cables with rubber (often called VIR) insulation in lieu of PVC have been widely used in the past and many of these are still in use. However, this type of cable has been almost entirely superseded by the PVC-insulated type which is cheaper and has

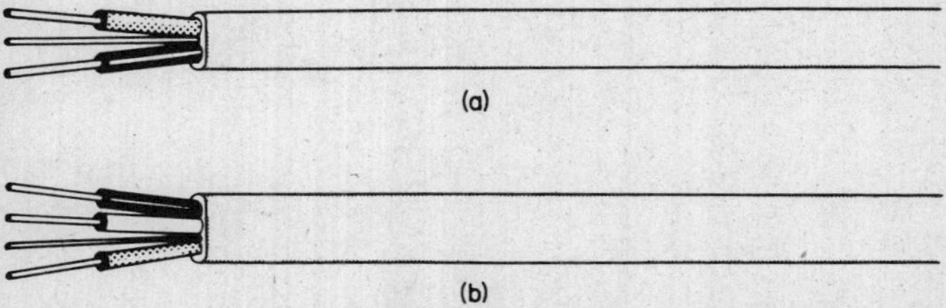

Fig. 16 *Flat PVC-insulated, PVC-sheathed cables: (a) twin and earth, (b) 3-core and earth (BICC Ltd.)*

other advantages. The lower cost is partly due to the fact that the copper conductors need not be tinned, whereas tinning is necessary with rubber insulation to prevent chemical action.

The information in this book is therefore based on system (a) above, i.e. PVC-insulated, PVC-sheathed cables as illustrated in Fig. 16. The cable for domestic work is usually twin and earth but three-core and earth may be required in some cases e.g. two-way switching.

The protective (earth) conductor *must* be taken to *all* outlet points, including lighting switches and lighting points as well as socket outlets.

In existing property, particularly the older type, the existing wiring may consist of single-core cables in lightweight, grip-type

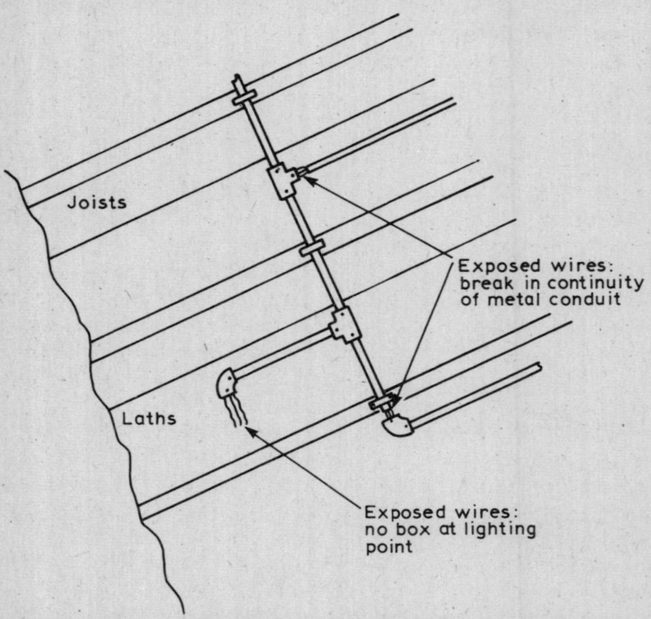

Fig. 17 *Typical existing conduit system showing exposed wiring*

conduit. This conduit does not usually provide satisfactory earth continuity and it frequently finishes short of the outlet points – see Fig. 17. When rewiring is carried out it is often convenient to use such conduits for cables feeding lighting switches or sockets so as to avoid chasing plastered walls. However, the cables must include a protective (earth) conductor and a rubber or PVC bush must be fitted at each end of the conduit to prevent damage to the cables. Provided PVC-sheathed cables are used, it does not matter if the conduits do not reach right up to the outlet points.

How are PVC-insulated, PVC-sheathed cables installed?

The cables are normally concealed in the building structure except in garages, workshops etc. and other areas where appearance is not particularly important, or in areas where the walls are bare brick.

When possible, cables should be run in roof spaces, below floors or above ceilings. Wiring along walls, such as drops to lighting switches or feeds to sockets, should normally be concealed in the plaster.

Cables in plaster. Cables run in plastered walls do not normally need mechanical protection; this would increase the cost and could cause problems in old buildings. However, in a new building it may be desirable to guard against damage during plastering, e.g. by the use of metallic or non-metallic conduit or channelling. Oval conduit may be preferable when the depth of plaster is limited. Short lengths of metallic conduit or channelling used in the above situation need not be earthed.

Cables buried in the plaster are subject to damage by nails etc. but the risk of this can be minimized by running all cables vertically or horizontally, *not* diagonally. The cable routes can then be deduced to some extent at a future date from the outlet positions. Concealed cables running horizontally should preferably be run at high level (within say 150 mm of the ceiling) or low level (above skirting level but within 300 mm of the floor).

Surface cables. It may not be possible to conceal cables running along walls in existing property as this would involve chasing the plaster and disturbing the decorations. However, provided the runs are carefully chosen, surface cables can be quite unobtrusive. Plastic mini-trunking can often be used to give a neat appearance or for mechanical protection when necessary. It is sometimes possible to run cables on the surface initially and to sink them into the plaster at a later date when redecorating.

Mini-trunking. This is made from white, high-impact PVC and may be screwed to the wall or, in suitable cases, fixed by an adhesive. It provides a very neat and robust installation without disturbing the decorations. A full range of accessories is available, including outlet boxes for switches, socket outlets etc. Kits are also available.

Cables under floors. Cables crossing joists under floorboards should pass through holes drilled in the joists. The cable should be at least 50 mm from the top or bottom of the joist, as appropriate – see Fig. 18. If this clearance is not possible the cables must be enclosed in metallic conduit or have equivalent mechanical protection at the crossing points. Joists should *not* be notched to receive the cables.

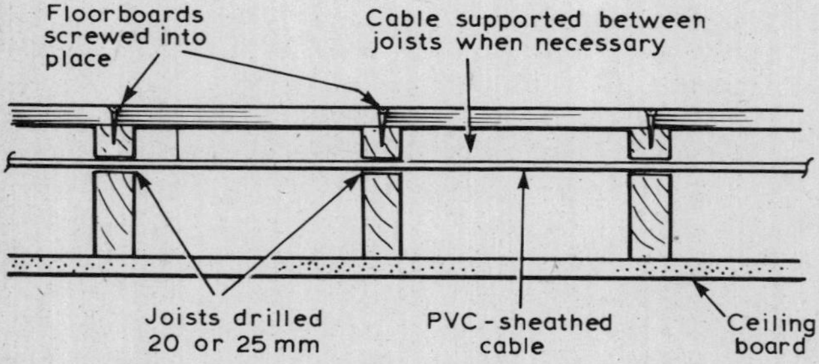

Fig. 18 *Installing PVC-sheathed cable under floorboards*

If the spacing between joists exceeds the maximum permissible spacing for the cable supports (see below) the cables must be supported on wood battens or by other suitable means. Cables running parallel to the joists within ceilings may rest on the ceiling.

After being lifted, floorboards should be screwed back rather than nailed and a short screwed section of floor board should be placed above each joint box and at other points where future access may be required.

Cable fixings. Cables should be fixed by suitable clips or saddles: typical fixings are shown in Fig. 19. The spacing between

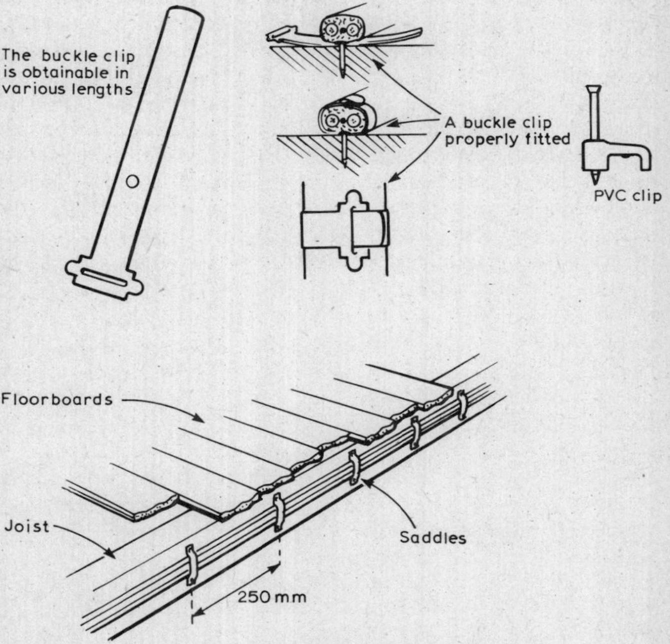

Fig. 19 *Clips and saddles for PVC-sheathed cables*

fixings should not exceed 400 mm for vertical runs and 250 mm for horizontal runs. Buckle clips have been widely used in the past but the PVC moulded type with single-hole fixing and steel pins are now very common. Both types are made in various sizes to match the cables concerned. When a number of cables run together, suitable saddles should be used.

Where cables are to be secured by clips or saddles, the run should first be marked out and the supports fixed at the correct spacing. The cable should then be laid across the fixings and pulled tight. The fixings should then be completed and tightened, working from one end of the run. Buckle clips should be lightly tapped with a hammer after being tightened.

Cables must be adequately supported in *all* cases, whether visible or not. There is sometimes a tendency to skimp on this, particularly for cables in the void below the ground floor where long loops of cable are sometimes left unsupported.

Protection against mechanical damage. Cables should be protected against mechanical damage when necessary, e.g. when run on the surface at low level. The protection may consist of metallic or plastic conduit, channelling or plastic mini-trunking.

Cable routes should be chosen so as to reduce the risk of damage, even though this may involve longer runs in some cases.

Protection may be necessary for cables in the roof space in areas where there is particular risk of damage, e.g. near the access trap or the water cistern. In such cases the cables should be run through, rather than on top of, the joists or enclosed in conduit or channelling.

Roof insulation. When thermal insulation in the roof space consist of granules of polystyrene there is a risk of chemical action between these and the PVC cable sheath. Prevent contact either by enclosing the cables in conduit or by other means.

Cables should not be in contact with, or be covered by, any type of thermal insulation; apart from the possibility of chemical action, such enclosure would reduce the permissible current rating of the cable.

Cables through walls. Cables passing through walls should be enclosed in a short length of metallic or plastic conduit with a bush at each end and the hole made good with incombustible material.

Contact with other services. Cables should not be in contact with gas, hot water or heating pipes.

Joint boxes. The use of joint boxes should be avoided as far as possible. If joint boxes are used they should be fixed and permanently accessible and, if under floorboards, an access trap should be provided in all cases.

Joints in PVC-sheathed cables must always be made in proper joint boxes with fixed screw terminals; taped joints should *never* be used.

Lath and plaster partitions. These are common in old property and it may be possible to run cables to lighting switches or sockets in the cavity. However, the run may be obstructed by cross members and it may be necessary to drill vertical holes in these.

If the switches, socket outlets etc. cannot be fitted on the structural members, suitable battens must be provided. Accessories should not be fixed to the laths.

Cables below ground floor. If the void is deep enough to permit crawling under the floor fairly easily, the cables should be clipped at the correct spacings to the sides or bottom of the floor joists.

If the void does not permit ready access, the cables may still be run in the void but should rest on the surface below. Lift floorboards at suitable points to obtain access to the void and use a 'fish' wire or a length of garden hose when necessary to pull in the cables.

In either case the cables should pass behind the skirting boards to outlet points, and care shold be taken that there is no strain on the cables at these points.

How are the cables routed?

In a normal two-storey house, the cables for the upstairs lights are usually run in the roof space, with vertical drops down the walls to the switches.

The wiring for the ground-floor lights and first-floor socket outlets is usually run in the space between the ground-floor ceiling and the first floor. The cables then rise behind the skirting boards to the first-floor sockets, and probably also to the immersion heater, and drop down the walls to the ground-floor lighting switches and possibly, to wall lights.

If there is a void below the ground floor, the ground-floor sockets, cooker etc. can usually be wired from below. If the ground floor is solid, the cables for the above items will have to be run either behind the skirting or horizontally along the walls, possibly in plastic trunking. Alternatively, the cables may be run in the ground-floor ceiling then dropping down the walls to the socket outlets etc. In some cases a mixture of the two methods may be the best solution.

Similar principles should be followed for wiring a bungalow.

What precautions should be taken when installing PVC-sheathed cables?

To prevent cables spiralling or kinking, the cable drum should be supported on a length of circular rod held clear of the floor so that the drum can revolve freely.

Avoid as far as possible the crossing of cables running on the surface or concealed in plaster.

PVC-sheathed cables may be stiff to handle in very cold weather and in such cases should be stored in a warm room before installation.

How are PVC-sheathed cables terminated?

The cable sheath should be removed for a suitable distance so that the sheath left on the cable protrudes just inside the outlet box or accessory. The cable cores should then be cut as necessary so that there will be a reasonable amount of slack, but not too much, and the insulation removed at the end of each core; care should be taken not to nick or damage the conductor since this could cause a breakage later or reduce the current-carrying capacity. The core insulation must reach right up to the terminal but must not be trapped inside or under it since this could prevent a connection being made.

If the hole in the terminal is much larger than the wire being used the bare conductor should be doubled back on itself so as to ensure an adequate contact; this can apply when a 13 A socket outlet is fed by a single cable.

All terminal screws should be well-tightened and the connections tested by pulling on each core.

There should be no tension or undue pressure on the cable cores when the accessory is fitted.

When the protective conductor is formed by a metal conduit or the sheath or armour of a cable (which is rarely the case in domestic work), the earth terminal of each socket outlet must be connected by a separate conductor to an earth terminal in the outlet box or enclosure; this is to ensure that the earth connection is not lost if the screws fixing the socket to the box are removed or become loose.

When PVC-sheathed cables are used, the protective conductor should be connected direct to the earth terminal on the socket or other accessory. Modern accessories usually include a metallic strip which automatically connects the fixing screws and outlet box to the earth terminal, in which case it is not necessary to provide a separate connection from the accessory earth terminal to the earth terminal on the outlet box. If this is not the case, or if there is any doubt, a separate connection should be made to earth the outlet box. The box must *not* be earthed by trapping the earth wire between the lid and the box.

All bare protective conductors inside outlet boxes or accessories should be enclosed in green-yellow PVC sleeving and any separate earth wires within the box should be either bare wire with sleeving as above or PVC-insulated wire coloured green-yellow.

As stated earlier, the protective conductor should be taken to *all* outlet points, including lighting switches and luminaires as well as socket-outlets.

The cable termination must be enclosed in an incombustible enclosure and when the accessory is mounted on the back of the cover plate the back must not be left open. The enclosure usually consists of a metal or moulded plastic outlet box or it may be incorporated in the accessory. Outlet boxes may be of the flush or surface-mounting type and for flush work the metal type is usual. Ideally the depth of the box should be slightly less than the depth of plaster; the box can then be mounted directly on the brickwork, without further chasing. However, this may not be possible with socket outlets since the plaster depth may be insufficient.

Suitable grommets or bushes must be fitted when PVC-sheathed cables enter metal boxes.

How is a plug connected to a flex?

This is a very common and important operation and is therefore described in detail. Fig. 20 refers.

1. Remove the plug cover by loosening the fixing screw and take out the fuse, if fitted. If the cord grip is held in place by screws, loosen them but leave them in place.
2. Lay the flex over the cord grip and measure the length of sheathing to be removed, usually about 50 mm. Then strip off the required length of sheathing and cut the three coloured cores so that they are the right length for connection to the terminals. On some plugs all the cores are the same length whilst on other types the lengths differ, the earth core being the longest.
3. Remove the insulation from the end of each core, about 15 mm, using a sharp knife or wire stripper and taking care not to nick the wire inside.
4. Place the flex in the cord grip and tighten the screws, if fitted. On some plugs the cord grip has no screws, the flex being held between two plastic grips which tighten when strain is placed on the flex.
5. Twist the bared ends on each core with thumb and finger so that there are no stray strands and then connect them to the plug terminals, first green-yellow to the earth terminal, then brown to the phase (marked L), and blue to the neutral (marked N). Some plugs have post-type terminals on which the wire passes through a hole and is held in position by a

screw and it may be necessary to double back the bared conductor to ensure an adequate connection, particularly when a small flex is being used. In other cases the terminals are the clamp type on which the wire is wrapped round a fixed terminal pin and held in position by a slotted screw. Wind the bared wire once round the terminal pin in a clockwise direction. In either case, make sure the insulation is taken right up to the terminal but it must not be trapped beneath it as this could prevent a connection being made. On some older appliances, the core colours in the flex may be red, black and green for the phase, neutral and earth respectively.

If the plug is being connected to a double-insulated appliance, such as a hair dryer, the flex will be two-core and should be connected to the phase and neutral terminals and there will not be an earth connection.

6. Insert a fuse of the correct rating for the appliance concerned, making sure that the clips hold the fuse tightly.

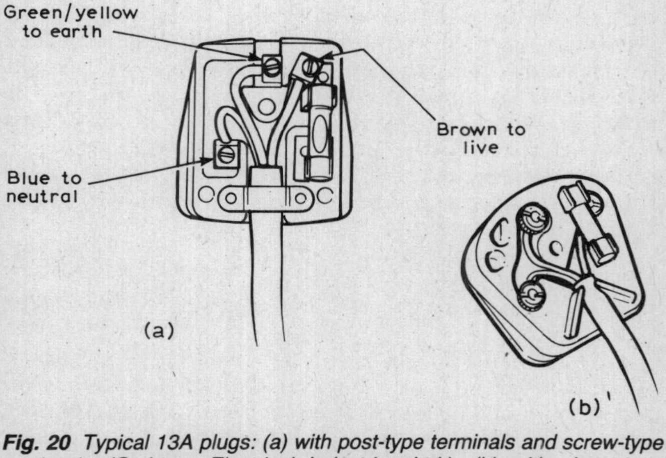

Fig. 20 *Typical 13A plugs: (a) with post-type terminals and screw-type cord grip (Crabtree Electrical Industries Ltd.), (b) with clamp-type terminals and screwless cord grip (MK Electric Ltd.)*

7. Before replacing the plug cover, check carefully that all terminals are quite tight, there are no stray strands of bare wire, the cable cores within the plug are slightly slack and not under tension, the cord grip is tight, the cores are connected to the correct terminals and the correct fuse is fitted. Then lay the cores in the grooves, fit the plug cover and tighten the fixing screw. Make sure that the plug cover fits tightly on to the base and the wire is not trapped between them.

The foregoing refers to 13 A fused plugs. The procedure for other types is generally similar except that the plugs are not usually fused.

How are cable cores identified?

PVC-sheathed cables. In twin and earth cables the red core should be used for the phase conductor, black for neutral and the uninsulated core for the protective conductor; this should be enclosed in green-yellow PVC sleeving where the cable sheath has been removed at terminations, i.e. within accessories, consumer units etc.

In some cases it may be necessary to connect the black core to the phase conductor, e.g. in a twin and earth cable from a ceiling rose to a lighting switch when the black core is used as the switch wire. The black core should then be enclosed in red PVC sleeving at each end.

Flexes. The colours used are brown for the phase conductor, blue for the neutral and green-yellow for the protective conductor. However, on older equipment it may be found that the colours are red, black and green respectively.

Single-core cables. These cables are normally used on a conduit installation and the colours should be red for the phase, black for neutral and green-yellow for the protective conductor. The switch wire between a ceiling rose and lighting switch should be red since it is connected to the phase conductor.

What is the loop-in system of wiring?

A system of wiring lighting, all connections being made at outlet points i.e. at switches, ceiling roses or luminaires mounted directly on ceilings or walls.

The arrangements differ according to whether the wiring consists of single-core cables in conduit or PVC-insulated, PVC-sheathed cables.

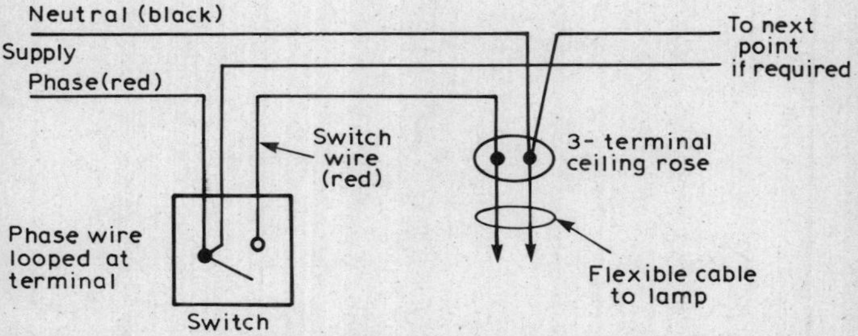

***Fig. 21** Loop-in system, using single-core cables in conduit. Protective conductor not shown; on metal conduit system this is normally the conduit*

Conduit system. Typical connections are shown in Fig. 21. The phase conductor is looped to all lighting switches and the neutral to all ceiling roses or luminaires; a further wire, called the switch wire, is run from the ceiling rose or luminaire to the switch. This method is not common on domestic work.

PVC-sheathed cable system. Connections for the loop-in system are shown in Fig. 22. Twin and earth cables are normally used but 3-core and earth cable may be required in some cases, e.g. 2-way switching. However, two twin and earth cables may be used in lieu of 3-core if desired.

On this system the phase and neutral conductors are looped to all ceiling roses and a further twin and earth cable is run from the ceiling rose to the switch.

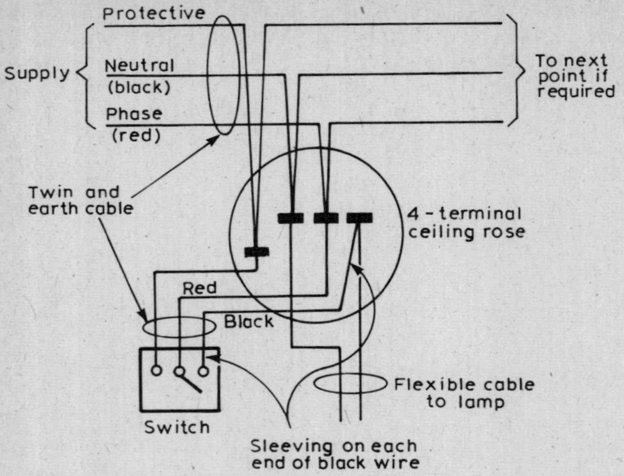

Fig. 22 *Loop-in system using flat twin and earth cables and 4-terminal ceiling roses*

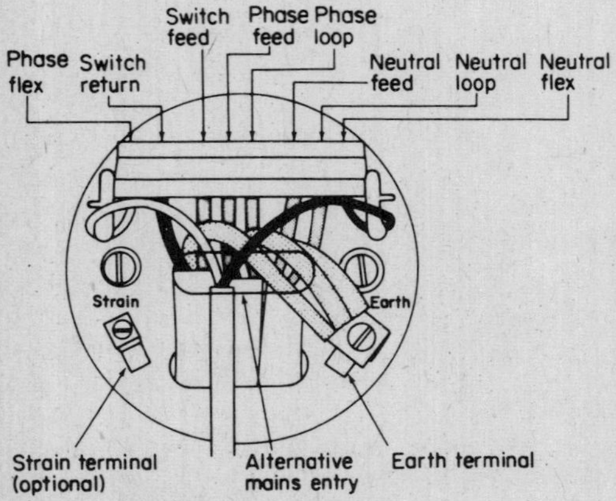

Fig. 23 *Typical ceiling rose*

The ceiling roses are of the 4-terminal type and may also include a terminal for a strain cord in the flexible cable. Fig. 23 shows a typical example.

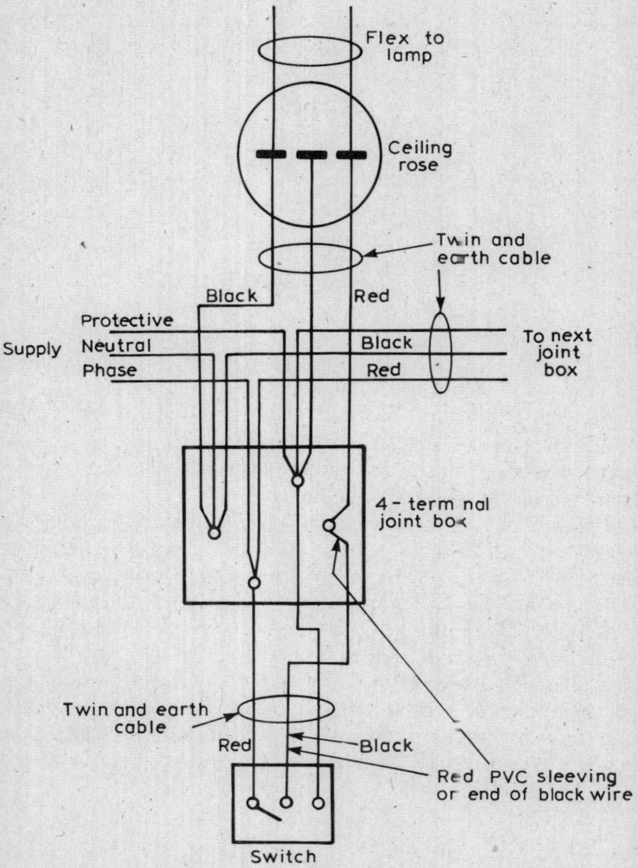

Fig. 24 Joint-box system using flat twin and earth cables

The full mains voltage is present at the ceiling roses whether the corresponding lighting switch is in the on or off position, and the phase terminal must therefore be shrouded to prevent accidental contact.

An advantage of the loop-in system is that all connections are readily accessible in the same room as the lights concerned and their locations are known, whereas joint boxes may be hidden under floors etc. which can cause difficulty when faults occur. Also, joint boxes may be a cause of faults and involve extra work in fixing and connecting. However, in practice, joint boxes are sometimes used in conjunction with the loop-in system in order to save cable when long runs are involved.

What is the joint-box system of wiring?

This system uses PVC-sheathed cables, fed via joint boxes which provide the connections for the mains supplies and the feeds to the ceiling roses or luminaires and switches. Fig. 24 shows typical connections.

The mains supply is looped to all the joint boxes, which are fitted at strategic positions so that the cables to the luminaires and switches are as short as possible.

The joint boxes are normally 4-terminal, which includes the protective conductor, but a 5- or 6-terminal type is sometimes used when the joint box feeds two or more luminaires or switches. The cables are normally twin and earth and the black switch wires should be identified by red PVC sleeving at all terminations, including joint boxes.

The main advantage of the joint-box system is that the length of cable used is kept to a minimum. However, the system has certain drawbacks as indicated earlier and, in general, joint boxes should be avoided as far as possible except in special cases.

4 Lighting

What is a luminaire?

This is the modern name for a lighting fitting. It includes all necessary parts on the fitting but not the lamps themselves. A batten lampholder, or a lampholder suspended by a flexible cord, is a luminaire.

How does a filament lamp work?

A normal filament (or incandescent) lamp consists of a fine wire, usually tungsten, within a sealed glass bulb filled with an inert gas

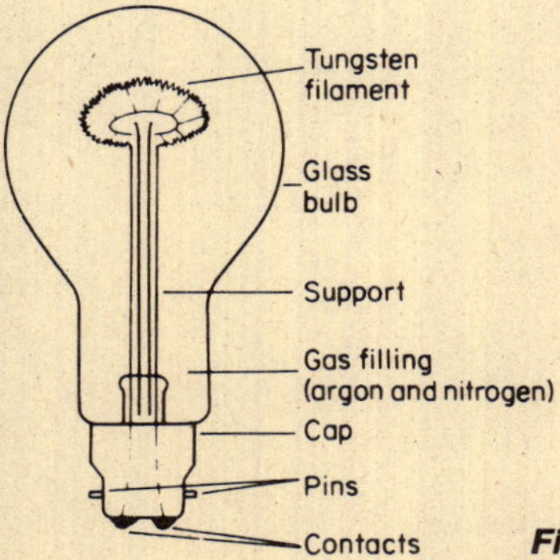

***Fig. 25** Typical filament lamp*

(Fig. 25). Light is emitted when the filament is heated to a high temperature.

Clear bulbs give a harsh light and should not be used in domestic situations unless enclosed within a luminaire.

Pearl lamps conceal the actual filament and give a softer light and should be used when the bulb is visible; their light output is virtually the same as the clear type.

On most modern lamps the filament is in the form of a spiral; and in the coiled-coil version the spiralled filament is again coiled upon itself; this increases the operating temperature and therefore the light output as compared with single-coil filaments.

'White' lamps give less shadow and softer lighting and are often preferred when the lamp is visible. Their light output is, however, about 10 per cent less than that of clear or pearl lamps. Mushroom lamps are of this type.

Lamp output sizes range from 15 W to 300 W but the maximum for domestic use is normally 150 W. Up to and including 150 W the lamps are normally bayonet type whilst higher ratings are usually the Edison screw type.

The normal life of a filament lamp is about 1000 hours. For maximum life the lamps should be used in the vertical 'cap-up' position; other positions, e.g. horizontal or cap-down, reduce the life slightly.

It is important that the lamp wattage used in a particular luminaire does not exceed the nominal rating of the fitting; otherwise overheating may result. This must be particularly watched with mushroom-type lamps, which are smaller than the equivalent plain or pearl lamps, and there may be a temptation to fit a 100 W or 150 W lamp into a 60 W fitting.

How does a fluorescent lamp work?

A fluorescent lamp consists of a glass tube containing a gas (usually argon) at low pressure and a little mercury and having an electrode at each end made of electron-emitting material. The inside of the tube is coated with a fluorescent material which produces light when current passes through the mercury vapour.

Control gear associated with the lamp includes a starting device and a ballast which limits the current through the lamp.

The nominal life of a fluorescent lamp is approximately 7000 hours.

How is a fluorescent lamp started?

There are two normal methods although the circuits used vary. These are (a) switch-start and (b) quick-start.

Switch-start. This is shown in Fig. 26. Initially the current flows through the ballast and the starter switch, which includes a bi-metallic strip; this heats up and after a short delay it operates,

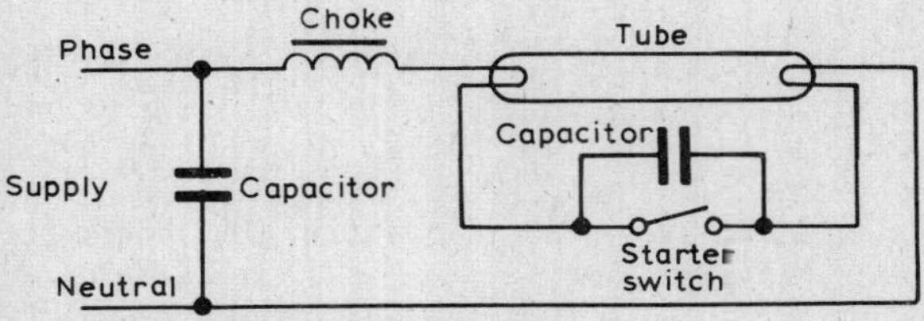

Fig. 26 *Typical switch-start circuit*

causing a momentary surge of high voltage between the lamp electrodes. This sets up the arc through the gas in the tube and this continues until the lamp is switched off. The delay period is necessary so that the electrodes are heated to the operating temperature by the time the starter switch operates.

Switch units for switch-start fittings are usually in canister form and can easily be replaced. The switch must match the size of lamp.

Switch-start lamps have the disadvantage that some delay and flickering occurs before the lamp is lighted.

Quick-start. Here, the fluorescent tube includes a metallic strip which is connected to earth, or alternatively the tube is located adjacent to earthed metalwork. A typical circuit is shown in Fig. 27.

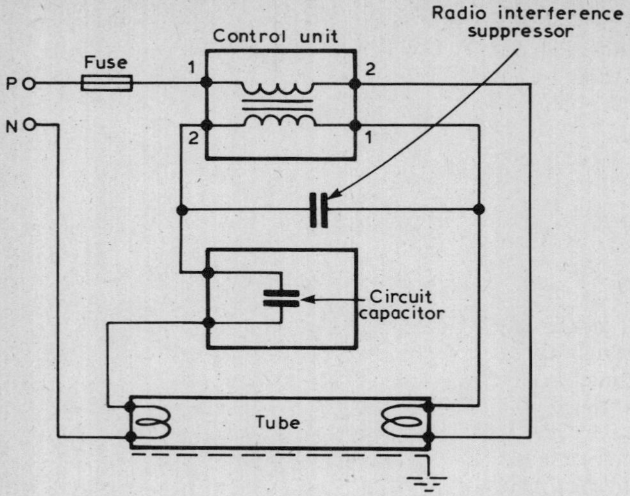

Fig. 27 *Typical quick-start circuit, semi-resonant type*

The lamp electrodes are heated via the transformer secondary and a starter switch is not required. When the arc is established the transformer acts as the current-limiting device. Although, as indicated above, an earthed metallic strip is required either on or adjacent to the lamp, there is, in fact, no electrical connection from the phase or neutral conductor to earth.

A switch-start lamp cannot be used in a quick-start fitting but either type of lamp may be used in a switch-start circuit.

What are the advantages of fluorescent lighting?

1. Longer lamp life, i.e. approximately 7000 hours compared with 1000 hours.
2. Higher lighting efficiency, e.g. one 85 W 'warm white' fluorescent lamp gives as much light as five 100 W filament lamps.

3. The light is of better quality, is more evenly distributed and a wider range of colours is available. However, the initial cost is usually greater for fluorescent than for filament lighting.

What types of fluorescent lamps are available?

Fluorescent lamps are available as (a) straight tubes in lengths from 450 to 2400 mm and ratings from 15 W to 125 W; (b) circular or U-shaped tubes in various sizes with ratings from 32 W to 60 W; and (c) miniature tubes in lengths of 150 mm to 525 mm with ratings of 4 W to 13 W.

Straight tubes are suitable for utility areas such as kitchens etc. and may be single or twin 1200, 1500 or 1800 mm in length depending on the room size, and should preferably be fitted with diffusers. Similar lamps without diffusers may be used in garages, workshops etc. or behind pelmets or in other situations where the tube is not visible. The 1800 mm size may be convenient behind long pelmets.

Circular fittings, preferably with diffusers, are used when a decorative effect is required, such as a lounge or dining room.

Miniature lamps are usually incorporated in other equipment such as cookers, refrigerators, mirror lights etc.

A wide range of colours is available, the most suitable for domestic use being white, warm-white or de-luxe warm-white. Other colours, e.g. pink, are available for special situations. Warm-white is normally considered best for general domestic use.

How are lighting circuits controlled?

Lights may be controlled by 1-way, 2-way or 2-way-and-intermediate switching. The switches are normally single-pole (SP) i.e. they only control one side of the circuit, and, as stated in Chapter 2, they *must* be connected in the phase conductor, *not* the neutral. Fig. 28 shows a typical arrangement with two luminaires controlled by a single 1-way switch.

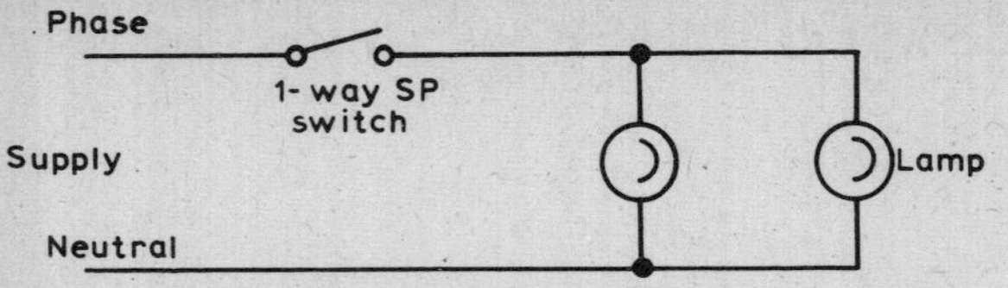

Fig. 28 Two luminaires controlled by one switch

Luminaires can also be controlled by a timeswitch although this is not very common in domestic accommodation except for security lighting.

Switches which are themselves visible and which control lighting which is not visible, e.g. in a roof space, should preferably include a pilot light to indicate the 'on' condition.

What is 2-way switching?

A 2-way switch has three terminals, one of which (called the 'common') is always connected to either of the other two (L1 and L2) with no 'off' position.

2-way switching is used when it is required to control one or more luminaires from two positions; a common case is a landing light in a 2-storey dwelling controlled by switches in the hall and on the landing, or a room with two doors.

Fig. 29*a* shows one method of connecting 2-way switches. With the switches in the positions shown the light will be on but if either switch is operated it will be extinguished; the further operation of either switch will bring the light on again. The arrangement in Fig. 29*a* is now normally used only when the wiring consists of single-core cables in conduit.

A method which is preferable when PVC-sheathed twin and earth cables are used is shown in Fig. 29*b*. It will be seen that the wiring from the light to the first 2-way switch is the same as for normal 1-way switching. This method can be used for new work or rewiring, or to convert existing 1-way switching to 2-way without disturbing the existing wiring. In the latter case, the

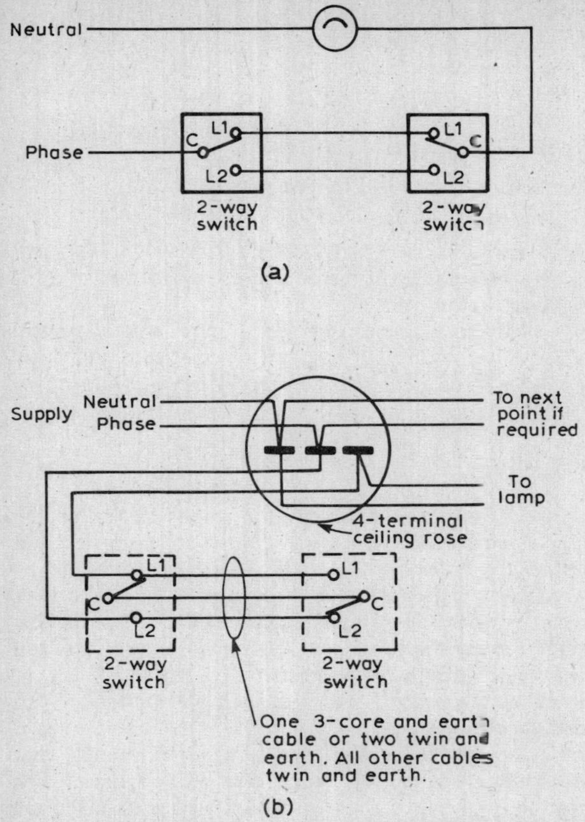

***Fig. 29** Two-way switching (a) using single-core cables, (b) using multi-core flat PVC-sheathed cables. Protective conductor not shown*

existing 1-way switch is replaced by a 2-way and a new 2-way switch fitted at the other position, with three wires between them connected as shown.

A two-way switch can of course be used as 1-way if desired; the common and one other terminal should be connected, the other terminal being unused. Some makes of cord-operated ceiling switches are only available as 2-way: also when one switch on a multi-gang switch is required to be 2-way, all switches are usually of this type.

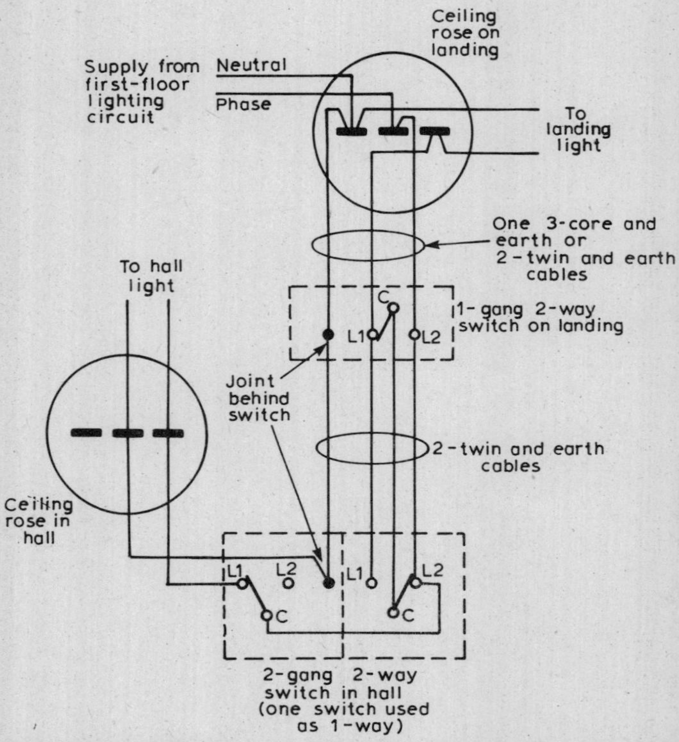

Fig. 30 *Wiring and connections for two-way switching on landing light and one-way switching on hall light. Protective conductor not shown*

2-way switching must not be confused with double-pole (DP) switching. 2-way switches are single-pole and must be connected in the phase conductor whereas DP switches break both the phase and neutral and are not normally used on lighting circuits.

How are hall and landing lights connected?

In a 2-storey dwelling it is quite common to have a 2-gang switch in the hall having a 1-way switch for the hall light and a 2-way switch for the landing light. In such a case it is essential that both the switches are on the same circuit, using a common phase conductor. It could be dangerous to take two separate phase connections, fed from separate circuits, to the switch unit since after removing a fuse, a person might wrongly assume the switch assembly to be dead. It is also essential that the neutral for both lights is taken from the same circuit.

Suitable connections are shown in Fig. 30. The phase and neutral connections for both lights are taken from the first-floor lighting circuit, an extra wire being fed from the landing to the hall light for the neutral; this should be jointed by suitable connectors at each switch position. It is quite common to find that the phase and neutral connections for 2-way switching have been taken from separate circuits; this is incorrect.

What is two-way-and-intermediate switching?

It is sometimes necessary to control luminaires from more than two positions e.g. a room with three entrances or a long corridor. This involves a 2-way switch at each end of the circuit and special intermediate switches in between. Fig. 31 shows the arrangement. Any number of intermediate switches may be fitted.

The intermediate switch reverses the connections of the two 'strappers' between the two 2-way switches, and thus performs the same function as the 2-way switches at the ends of the circuit; i.e. in Fig. 31, in one position the intermediate switch connects terminal L1A to L2A and L1B to L2B whilst in the other position L1A is connected to L2B and L1B to L2A.

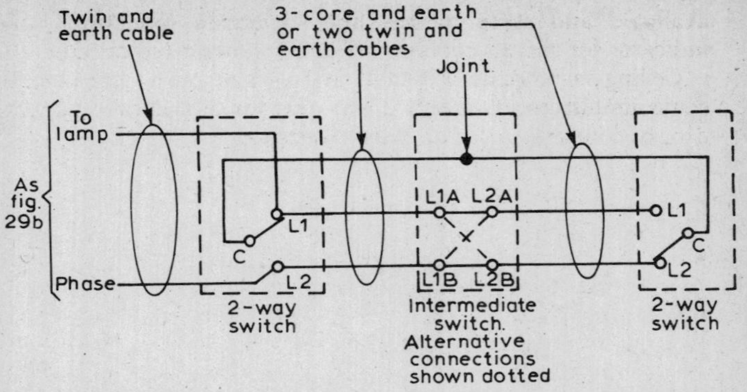

Fig. 31 *Connections for two-way and intermediate switching. Protective conductor not shown*

When are cord-operated ceiling switches used?

Ceiling switches are cord-operated, each operation changing the condition of the switch, i.e. 'off' to 'on' or 'on' to 'off'.

They are often used in bathrooms or similar situations where a wall-mounted switch is not permissible (see Chapter 2). They are also commonly used in bedrooms so that the light can be switched on or off from the bed; in such cases 2-way switching is often used with a wall-mounted switch at the door and a cord-operated switch over the bed.

A ceiling switch can also be convenient when the wiring to a wall-mounted switch would present problems e.g. when partition walls are not full height, or when a wall-mounted switch would involve chasing the plaster and disturbing the decorations; this situation often arises on rewiring projects. A ceiling switch can usually be wired quite easily in the roof void or ceiling space.

It is not possible to tell by inspection whether a ceiling switch is in the on or off position; thus if the appliance concerned is not luminous it is preferable for the switch to include a pilot light which glows when the switch is on. Such switches are readily available and some models also include a mechanical flag indicator for the off condition. Fig. 32 shows a typical unit.

Ceiling switches are available in 1-way or 2-way versions but some manufacturers supply 2-way only for certain ratings; these can, of course, be used as 1-way if desired.

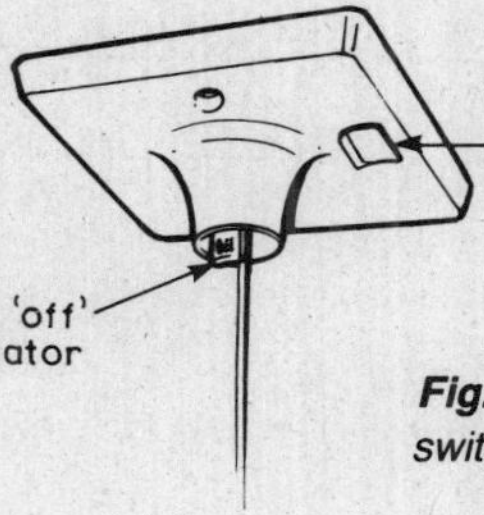

Fig. 32 Typical cord-operated ceiling switch (Ashley Accessories Ltd.)

Cord-operated ceiling switches may also be used to control immersion heaters or shower units – see Chapter 5. Ratings of up to 45A are available.

How can a cord-operated ceiling switch be added to an existing bedroom light?

In cases where a bedroom light is controlled by a 1-way wall-mounted switch at the door and it is desired to provide additional control from the bed, the ideal method is to provide 2-way switching, using a new 2-way switch over the bed connected as shown in Fig. 29.

However, this requires additional wiring to the wall-mounted switch, which may involve chasing the plaster and disturbing the decorations. If this cannot be accepted, an alternative method, providing limited control, is to connect the new ceiling switch over the bed as 1-way in series with the existing wall-mounted switch. The existing circuit should be intercepted at a convenient point in the roof space and a joint box fitted from which a 1 mm^2

twin and earth PVC-sheathed cable should be run to the new ceiling switch.

This arrangement permits the light to be switched on or off from the bed, provided the wall-mounted switch is 'on', but it is necessary each morning to reset the switches with the ceiling switch 'on' and the wall switch 'off'.

The above is not quite so convenient as 2-way switching but it is a useful compromise.

A further possibility, which provides 2-way switching without chasing the plaster, is to abandon the wall-mounted switch and replace it by a new 2-way cord-operated ceiling switch near the door and a new 2-way ceiling switch over the bed, wired as shown in Fig. 29.

How is dimming of lighting carried out?

A dimmer permits the lighting level to be varied from full to off and is used mainly in bedrooms, lounges, halls etc. and for night lights. Modern dimmers consist of solid-state devices which consume a negligible amount of electricity. Thus dimming reduces the electrical consumption although this is usually a secondary consideration.

Fig. 33 *Typical dimmer units: (a) combined dimmer/on-off switch (MK Electric Ltd), (b) two-gang dimmer (Home Automation Ltd.)*

With *filament lighting* the dimmer switch can simply be fitted in lieu of the normal on/off switch without any modification to the wiring and it will normally fit on the same standard outlet box.

Dimmers usually include a separate on/off switch and a local fuse. Units with 2-way switching are available but these can be used as 1-way if desired. Fig. 33 shows typical units. Multi-gang dimmers are also available for use when two or more groups of lights are to be controlled separately from the same position.

For dimming portable lights, a dimmer/adaptor is available which plugs into a standard 13A socket outlet and into which two portable lamps can be plugged. Another type can be wired into the flexible cable supplying the portable lamp.

On some dimmers the lower lighting level can be pre-set and can then be achieved by flicking a switch, without further adjustment.

All dimmers have maximum and minimum ratings in watts, and these must be complied with. Common values for domestic use are 600 W maximum and 40 W minimum but the figures vary for different models. It may be necessary to change a low-wattage lamp, e.g. 25 W, to a larger type if dimming is required.

Dimming *fluorescent lighting* is not so simple. Modifications to the luminaires are needed, as is extra wiring between the dimmers and the lighting fittings. Dimmers for filament lighting are not suitable for fluorescent lights and mixed loads of the two types of lighting cannot be controlled by the same dimmer. If dimming is required on a domestic installation which has fluorescent lighting, it is preferable and usual to provide some filament lighting in the same room and apply dimming to this, the fluorescent lights being switched off.

Can lighting be controlled remotely?

Systems are available which permit lights to be dimmed or switched on and off from a remote position e.g. from an armchair or bed; this is especially helpful to an invalid or disabled person.

The hand control is a compact device which operates the

receiver and this, in turn, controls the lights. The receiver may be wall- or ceiling-mounted or combined with a ceiling rose.

The system uses infra-red rays and the range is about 10 m provided the control unit and receiver are in line of sight; the system will operate through clear glass.

How are time-delay switches used?

A time-delay switch, when operated, connects the circuit for a pre-determined period, after which it is automatically disconnected.

These switches can be used to control lighting, e.g. on leaving the house an outside light may be left on for a short period so as to enable a person to reach another lighted area or a car or garage.

They may also be used for controlling hall or landing lights so that they are not left on unnecessarily for long periods. However, they should be used with discretion, particularly when staircase lighting is concerned, so as to avoid danger if the area is suddenly plunged into darkness. In such cases a fairly long delay period should apply and the lights should preferably be dimmed gradually before being extinguished.

The switches are usually of the push type and fit on to a standard outlet box. The push button restores to normal on expiry of the delay period. 1-way, 2-way and intermediate types are available with delay periods of up to 20 minutes. The delay period can be varied by means of a screw which is normally hidden behind the cover plate.

How are wall lights wired?

Wall lights may be connected to a convenient lighting circuit, a ring or radial circuit feeding 13 A socket outlets, or a separate 5 A way on the consumer unit. On a new installation it is usual to connect them on a lighting circuit, in which case they are usually wired from above. However, when they are being added to an existing installation or during rewiring it is often easier to connect them to a socket outlet circuit in the same room; this may reduce the chasing of plastered walls and also avoid the need to lift carpets and floorboards and move furniture on the floor above.

In connecting to a socket outlet circuit it is preferable to loop from the terminals of a convenient socket to a fused connection unit, fused at 3 A, using 2.5 mm^2 twin and earth cable. The connection unit may serve as the lighting switch if fitted in a suitable position, or alternatively separate switches can be used.

The cable from the connection unit to the lights should be 1 mm^2 twin and earth; this size should also be used if the lights are connected to a lighting circuit and in this case no further fusing is necessary. One fused connection unit may serve a number of lights.

In some cases a switch is provided at the door of the room, either in addition to or in lieu of local switches, thus permitting all the lights in the room to be controlled from the door.

The protective conductor must be taken to the wall lights in *all* cases and any metalwork on the fittings connected to it.

How are lighting circuits arranged?

The IEE Wiring Regulations recommend that, in assessing the load on a lighting circuit, the actual loads should be used, if known; otherwise a minimum of 100 W *per lampholder* should be assumed.

In domestic premises it is usual to use 5 A circuits for lighting and at 240 V this is equivalent to 1200 W; thus a circuit with 12 lamps is theoretically possible, fewer if some lamps exceed 100 W. However, in practice, the number of *lighting points* on a circuit should not exceed eight (a luminaire with more than one lamp being counted as one point). This figure allows for multi-lamp fittings, gives scope for extra lights at a future date, and minimises the inconvenience if a circuit fails.

In a normal 2-storey house it is good practice, and usual, to have a lighting circuit for each floor, subject to the above. In large bungalows or flats two circuits should still be used but in very small properties the decision will depend on the number of lights and individual preference. If a lighting circuit fails, temporary lighting can of course be obtained from portable lamps via socket outlets.

The cable size can be 1 mm^2 whichever system of wiring is used, PVC-sheathed twin and earth being the most common.

How are filament lamps connected to the fixed wiring?

Luminaires with filament lamps may be suspended from a ceiling rose by means of a flex pendant or be close-mounted direct on to a wall or ceiling.

Pendant type. Flex pendants should preferably be of the circular sheathed type rather than twisted unsheathed and should be three-core if the luminaire includes any metalwork or brass lampholders. Modern brass lampholders include an earth terminal and any existing ones without it should be changed to a modern brass type or replaced by the plastic type which does not require earthing.

The flex should be held by the cord grips or anchorage slots in the ceiling rose and lampholder so that there is no strain on the terminations and the flex sheath should protrude inside the ceiling rose and lampholder, with no unsheathed sections outside.

Close-fitting luminaires. The simplest type of wall or ceiling-mounted fitting is a batten lampholder and for lamps of not more than 60 W which are fed by a single cable the fixed wiring may be taken direct to the lampholder terminals. If the lamp exceeds 60 W the lampholder should be of the type on which the fixed wiring is terminated on separate terminals, with heat-resisting wiring from these to the actual lampholder; this type should also be used when two or more cables are being looped at the lampholder.

For other types of wall or ceiling-mounted luminaires the requirements are similar, i.e. up to 60 W and with a single cable, the wiring may terminate directly on the fitting. For more than 60 W, or when cables are being looped at the fitting, the fixed wiring should terminate at a joint box fitted behind the luminaire with heat-resisting flex to the lampholder.

The above requirements regarding the use of HR flexible cable are most important since normal wiring will quickly become brittle and disintegrate if subjected to overheating.

Luminaires with *fluorescent* lamps may be connected directly to the fixed wiring and the above precautions are not necessary.

5

Power

What are the requirements for socket outlets?

A socket outlet provides a means of connecting portable appliances to the fixed wiring.

In the UK the standard socket outlet at the present time is the flat-pin type for use with fused plugs having a maximum rating of 13 A. The socket outlets and plugs are to BS 1363.

Other types of socket outlet, including 2, 5 and 15 A with round pins, and other non-standard types, are not now installed (except possibly on existing systems), although many are still in use.

Socket outlets may be single or twin and switched or unswitched. In general, the switched type is preferable except possibly in halls, corridors, landings etc.

Switched sockets normally have single-pole (SP) switching but sockets with double-pole (DP) switching are available; the wiring is the same in either case. SP switching is normally adequate. It is most important that the SP switch is in the phase conductor. See Chapter 2.

Socket outlets should be fitted and connected so that the phase socket (marked 'L') is on the right and the neutral (marked 'N') on the left when viewed from the front, and the plug should be wired to correspond.

Switched socket outlets with indicating lights are available and are useful for supplying non-luminous appliances such as freezers or heaters. Alternatively a plug with an indicating light may be used.

Special socket outlets are also available for use with clocks, shavers etc., and these are described later in this chapter.

What are the requirements for fused plugs?

Fused plugs for use with the 13 A sockets described above should be fitted with fuses to BS 1362. Fuse ratings of 2, 3, 5, 10 and 13 A are available, coloured black for 2, 5 and 10 A, red for 3 A and brown for 13 A. The 3 A and 13 A are normally used, and are the most widely available. A 3 A fuse is suitable for loads of up to 720 W and the 13 A size will carry up to 3 kW. The approximate consumption for various appliances is given in Table 1, page 5.

However, some appliances require a larger fuse than indicated by their nominal loading due to an initial surge when switched on and in such cases the manufacturer's recommendations should be followed. Items in this category include colour television sets, freezers, spin dryers and other items which include an electric motor.

Care should be taken that the current rating of the flexible cable is at least equal to that of the fuse in the plug. Table 3 (page 115) gives the current ratings of flexes.

As stated on page 77, an immersion heater of 3 kW rating or more should not be connected via a 13A plug and socket.

Plugs incorporating a neon indicating lamp are available and these can be used to supply non-luminous loads such as freezers etc.

It is important that fuses in 13 A plugs are of the correct rating and type and that they are labelled 'BS 1362'. Fuses intended for other types of fused plug will not fit correctly into 13 A plugs and may cause overheating

How are socket outlets fitted?

When mounted on a vertical wall or other surface, socket outlets should be *at least* 150 mm above floor or worktop level, measured to the bottom of the socket; this avoids damage to the plug or flexible cable. Socket outlets should never be fitted on skirting boards as they could be damaged by cleaning appliances or heavy furniture. If the socket outlets will be used by disabled or elderly persons it may be preferable to fit them at a height of approximately 1 m.

The colours to be used for the fixed wiring to sockets and for the flex from the plug to the appliance are given in Chapter 3.

Socket outlets must be enclosed in metal or insulated outlet boxes as described in Chapter 3.

How can two single socket outlets be fitted on the same outlet box?

Boxes are available on which two single sockets can be fitted although the appearance is not quite so neat as that of a twin socket. Fig. 34 shows typical outlet boxes.

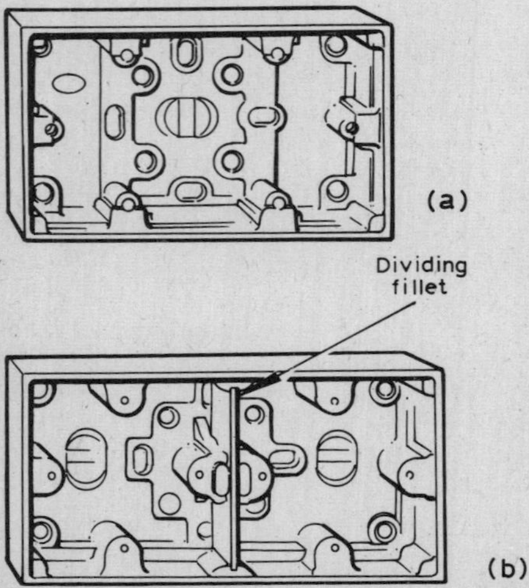

Fig. 34 *Outlet boxes: (a) two-gang moulded surface mounting, (b) dual-moulded surface mounting (MK Electric Ltd.)*

What special precautions are necessary for socket outlets which may feed portable appliances outdoors?

In accordance with the IEE Wiring Regulations (Reg. 471–12), such socket outlets must be protected by a residual-current device having an operating current not exceeding 30 mA. See Chapter 1.

What is a socket outlet ring circuit?

The object of the ring system for feeding socket outlets is to facilitate a generous provision of sockets with the minimum amount of cabling and protective equipment, advantage being taken of the fact that all the socket outlets will not be in use at the same time.

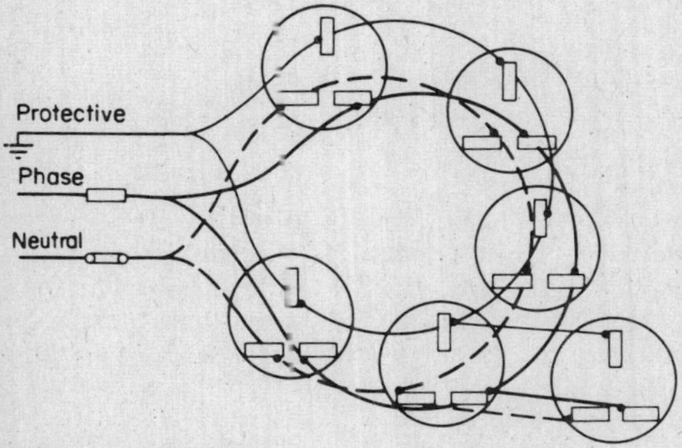

Fig. 35 *Ring circuit for socket outlets*

A socket outlet ring circuit, as used with 13 A plugs and sockets to BS 1363, commences at a 30 A way in the consumer unit or distribution board and is looped to the terminals of all outlets on the ring, i.e. sockets, fused connection units or joint boxes, and returns to the consumer unit. See Fig. 35. The two phase conductors are connected to the same fuse or MCB and the two neutrals and protective conductors are also connected to the appropriate terminals in the consumer unit. Thus all outlets on the ring are fed from two directions.

The cable on domestic installations is normally $2.5\,mm^2$ flat, twin and earth, PVC-sheathed.

The number of socket outlets and fixed appliances connected to the ring is unlimited but the floor area served by a ring must not exceed $100\,m^2$. The anticipated load must also be taken into account, particularly in kitchens, and must not exceed 7.2 kW.

It is usual and very desirable to have two separate rings in a dwelling except in very small properties.

Spurs may be connected to ring circuits and they may be fused or non-fused as described later (pages 72–73).

Fixed appliances fed via ring circuits may be connected via a plug and socket outlet or a fused connection unit (see page 71). The latter course is preferable since, if a socket outlet is used, this may be used for other purposes and the fixed appliance left disconnected.

What is a radial circuit?

A radial circuit, as used with 13 A plugs and socket outlets to BS 1363, is similar to the ring circuit described above except that only one set of wires (phase, neutral and protective conductor) is connected at the consumer unit and the circuit finishes at the last outlet. See Fig. 36. The permissible floor areas served also differ from those for a ring circuit.

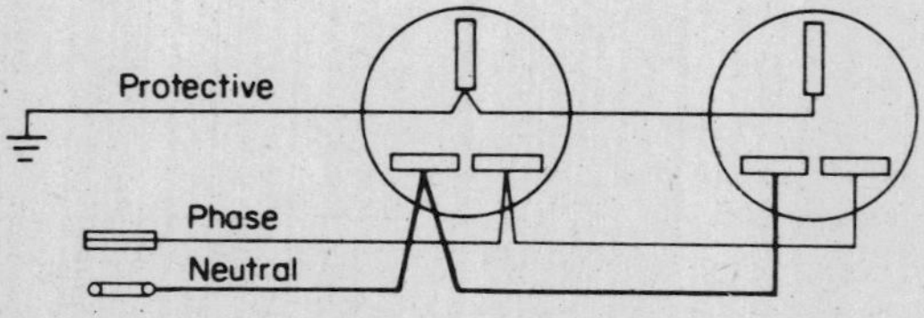

Fig. 36 *Radial circuit for socket outlets*

Radial circuits may be rated at 20 A or 30 A, as described below:

20 A radial circuit. A radial circuit fed via a 20 A MCB or cartridge or rewirable fuse may feed an unlimited number of outlets provided the floor area served does not exceed 20 m^2 and the anticipated load does not exceed 4.8 kW. The cable size should be 2.5 mm^2 twin and earth flat PVC-sheathed.

30 A radial circuit. A circuit fed via a 30 A MCB or *cartridge* fuse (not a rewirable) may feed an unlimited number of outlets provided the floor area does not exceed 50 m^2 and the anticipated load is not more than 7.2 kW. The cable size in this case should be 4 mm^2.

What are fused connection units?

These are used to connect fixed appliances, such as wall-mounted fires, hand dryers, extractor fans etc., directly to ring or radial circuits feeding 13 A socket outlets, i.e. a plug and socket are not used.

Fig. 37 *Typical fused connection units: (left) with flex outlet, (right) with indicating light*

The maximum fuse rating is 13 A and the fuse rating must not exceed that of the cable supplied via the fuse. The fuses are to BS 1362 and are of the same type and ratings as are used in 13 A plugs – see page 67.

The units are normally square and fit on to standard outlet boxes. They are available in various types: with or without indicating lights, switched or unswitched, single or double-pole switching, surface- or flush-mounting, metal or insulated cover plates, and with or without flex outlet. Fig. 37 shows typical units.

It should be noted that DP switching is required if the fused connection unit controls an appliance fitted with heating elements which can be touched; and the unit must be in a readily-accessible position. In this connection, the sheath of a silica-glass element is regarded as part of the element.

Fused connection units may also be used to feed isolated luminaires, e.g. wall-mounted, when the connection of these to normal lighting circuits would involve long cable runs or other difficulties. See Chapter 4.

What is a fused spur?

A fused spur is connected to a ring or radial circuit via a fused connection unit. The rating of the fuse in the connection unit must not exceed that of the cable forming the spur and in any event must not be more than 13 A.

A fused spur can be used to feed socket outlets, in which case the conductor size for PVC-insulated cable must be at least 1.5 mm^2; however, this situation is not very common as any socket outlets not connected in the actual ring circuit are normally fed by *unfused* spurs – see below.

When a fused spur is used it is not necessary to provide additional fusing for fixed appliances at the termination of the spur. However, a local switch will usually be necessary adjacent to the appliance unless the fused connection unit feeding the spur is reasonably adjacent to it.

The number of fused spurs on a circuit is unlimited, subject to the anticipated load and the maximum floor areas quoted earlier.

What is a non-fused spur?

A non-fused spur is connected directly, i.e. without fusing, to the ring or radial circuit. The connection may be made at a socket outlet or joint box or at the origin of the circuit in the consumer unit. The cable size for a non-fused spur should be the same as for the ring or radial circuit to which it is connected.

A non-fused spur may feed only one single or one twin socket outlet or one fixed appliance. When a non-fused spur feeds an item other than a socket outlet, local fusing, via a fused connection unit, must be provided at the termination of the spur.

A non-fused spur is often convenient and saves cable when used to feed a socket outlet or other appliance situated some distance from the ring or radial circuit, since only one cable needs to be run to the remote point.

The number of non-fused spurs on a circuit must not exceed the total number of socket outlets and fixed appliances connected directly to the circuit; in this connection, a twin socket outlet should be counted as two outlets.

How are fixed appliances connected?

Normally by permanent wiring and not via socket outlets.

Appliances with loadings of more than 3 kW, and any which involve a 3 kW load for long periods, such as immersion heaters, should be on individual circuits from the consumer unit and not connected to ring or radial circuits feeding socket outlets. In such cases each appliance should be controlled by a separate switch, fitted adjacent; this need not normally include a fuse since the circuit will be fused at the consumer unit.

Smaller appliances, such as fixed fires, small water heaters (less than 3 kW) hand dryers, small cookers, fans etc., may be connected to ring or radial circuits feeding socket outlets. Local fusing is essential and the appliances should be connected via fused connection units – see page 71.

Items with very small loadings, such as clocks, shaver units, domestic fans etc. may be connected either on ring or radial circuits as described above or directly to 5 A lighting circuits; in the latter case, local fusing is not necessary.

Combined heat-and-light units must *not* be connected on lighting circuits. See page 91.

Special requirements applicable to rooms containing a fixed bath or shower are detailed in Chapter 2.

How is an immersion heater connected?

An immersion heater rated at 3 kW or more should be on a separate circuit from the consumer unit.

The most common loading for domestic use is 3 kW and the fuse or MCB in the consumer unit should be 15 A or 20 A. The cable should be 2.5 mm^2 twin and earth, flat, PVC-sheathed.

The immersion heater should be controlled by a 20 A DP switch fitted reasonably adjacent. However, as stated in Chapter 2, it is not permissible to fit a wall-mounted switch in a position normally accessible to a person using a fixed bath or shower and this may preclude a wall-mounted switch in the bathroom or airing cupboard. In such cases it is necessary to fit either a wall-mounted switch outside the room or a cord-operated ceiling switch inside; in either case the switch should be 15 A or 20 A DP and should preferably include an indicating light.

It is often convenient to fit a further switch in a suitable position, such as the kitchen; the supply cable should then be run from the consumer unit to the first switch and then to the switch or junction box adjacent to the immersion heater.

The final connection to the heater should be 2.5 mm^2 three-core heat-resisting flex; this can run direct from the switch, if adjacent; otherwise a junction box with flex outlet plate should be fitted and the flex run from this.

The brown phase core in the flexible cable should be connected to one terminal of the thermostat and the other side of the thermostat to the heater.

When a single immersion heater is used, it is usual to fit this into the top of the cylinder in a vertical position. This provides a small amount of hot water quickly and heats almost all the water in the cylinder provided the element reaches almost to the bottom.

In some cases two separate heaters are fitted, one short (usually 2 kW) at the top and another, fitted horizontally near the bottom. The lower heater will, in time, heat almost all the water in the cylinder but will not provide hot water quickly. The top heater provides a small amount of hot water quickly. For such cases a special switch is available (Figs. 38 and 39) which includes a main DP 'on/off' switch, which disconnects both heaters, and a

Fig. 38 *Typical immersion heater switches (a) for single heater, (b) for dual heater (MK Electric Ltd.)*

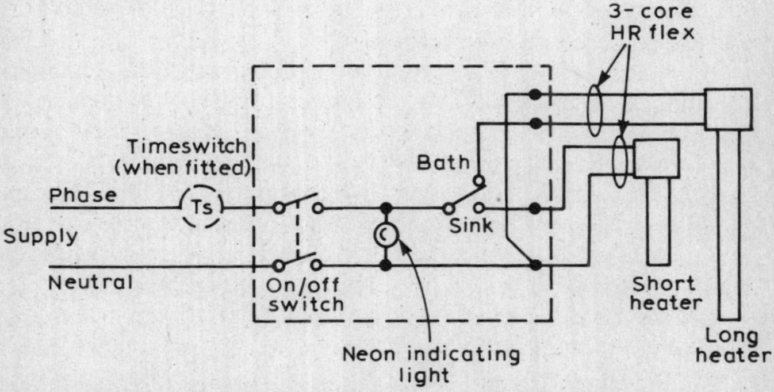

Fig. 39 *Connections for dual immersion heater. (If selector switch is not adjacent to heaters, separate isolating switches are required adjacent to heaters)*

changeover switch which connects the supply to either of the two heaters but not both at once. This switch may be fitted near the hot water cylinder when permissible (see above) or remote, e.g. in the kitchen; in this case an extra wire is required from the switch to the cylinder and a local DP isolator is required adjacent to each heater as before. See Fig. 39.

Dual heaters are also available, consisting of one short and one long element fitted in the same housing, each with its own thermostat (Fig. 40). The heater is fitted vertically in the top of

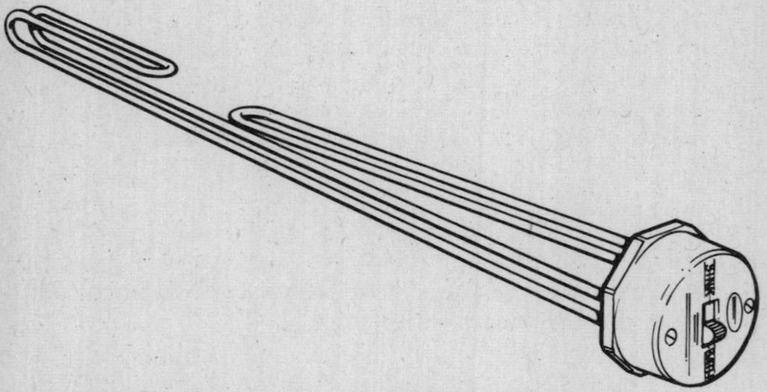

Fig. 40 *Dual immersion heater showing selector switch (Santon Ltd.)*

the cylinder and functions in the same manner as two separate heaters. On some models the two heaters must be wired away to a separate 'dual' switch as described above whilst others, as in Fig. 40, include an integral selector switch in the top of the heater which permits either, but not both, to be used at once; in this case only a normal twin and earth supply cable is required. However, the two heaters may still be wired out separately to a remote control switch if desired, the switch on the heater then being by-passed.

A dual heater with built-in selector switch must not be fitted in such a position that the switch is normally accessible to a person using a fixed bath or shower (see Chapter 2) and if this condition cannot be met, the selector switch must be disconnected and the heaters wired away to a switch in an acceptable position as indicated above.

Various other switching arrangements are possible e.g. with two heaters it is sometimes convenient for only one of them to be switched remotely, e.g. from the kitchen, the other being controlled locally. It is also possible to wire the circuit so that both heaters can be used at the same time; this involves either a 30 A circuit for the two heaters or two separate 15 or 20 A circuits.

If you intend to install a dual heater and to use remote switching, not the selector switch on the heater, you should check before purchase that the selector switch *can* be by-passed.

Can I connect a 3 kW immersion heater via a 13 A plug and socket?

This is not advisable. Apart from the fact that a socket outlet must not be fitted in a bathroom, the plug and socket carrying this load for prolonged periods may overheat when fitted near a hot water cylinder in a confined space.

How should the thermostat be set?

Immersion heaters should always be controlled by thermostats; otherwise boiling may occur if you forget to switch off. On some models, particularly the dual type, the thermostat settings are fixed and cannot be varied. This point should be carefully checked when buying the heater if you wish to be able to vary the settings yourself.

A good average setting is 60°C, with a maximum of 70°C in soft-water areas. A high temperature increases standing losses but gives greater storage capacity. However, in hard-water areas the setting should not exceed 60°C so as to limit the rate of scaling.

When two heaters are fitted, or a dual type, the thermostat for the lower (or longer) heater should be set at 10–20°C higher than that of the top element; otherwise the longer element will cut off too soon or may not even switch on. The best settings will be found by experience but expert advice should be obtained when necessary.

What about scaling?

Immersion heaters with a special finish are available for hard-water areas since hard water reduces the life and efficiency of normal heaters. In such cases, consult the manufacturers.

A long immersion heater operates at a lower surface temperature than a shorter heater of the same electrical loading since it has a greater surface area. Heaters should therefore be as long as possible, particularly in hard-water areas; this reduces scaling and increases the life of the heaters. Ideally, a top-entry heater should reach to within about 75 mm of the bottom of the cylinder.

How can an immersion heater be timeswitch-controlled?

It is often convenient and economical to connect an immersion heater via a timeswitch. A common arrangement is to switch on the heater for an hour or two early in the morning and also, perhaps, in the late afternoon.

The timeswitch should include a by-pass switch (sometimes called an 'anticipatory' switch) or, if not, a separate by-pass switch should be fitted so that the heater can be used at *any* time, if desired. If two immersion heaters are fitted, connect the timeswitch in the common supply as shown in Fig. 39 so that either heater can be used. The timeswitch should be suitable for a 20 A load and should be of the permanently wired type, not the type which plugs into a 13 A socket outlet.

When the off-peak tariff is in use it may be economical to heat the water at night at the cheap rate. If the heater is permanently connected to the *off-peak* consumer unit a separate timeswitch need not be provided but the water cannot then be heated outside the off-peak period. If this is inconvenient, connect the heater on the *normal* supply and fit your own separate timeswitch; you can then set this to bring on the heater during the off-peak period (i.e. the cheap rate) but by using the by-pass switch you can use it at other times if you wish.

The use of timeswitches is covered in more detail on pages 89–91.

Why is thermal lagging important?

It is obviously desirable that heat loss from the hot-water cylinder should be kept to a minimum so as to reduce the electrical consumption and therefore the cost. The cylinder should therefore be thermally insulated, preferably using a purpose-made jacket of fibreglass or similar material.

The minimum thickness should be 100 mm and it is claimed that a 130-litre cylinder of water, heated overnight to 60°C, loses only 6°C in 12 hours if a lagging jacket of this thickness is fitted.

Lagging is particularly important when the water is heated on the off-peak tariff, since the water may not be drawn for several hours.

How is an immersion heater fitted?

Some hot water cylinders have a flange already fitted in the top for an immersion heater. If not, a hole must be cut and a flange fitted; this requires special tools and skill and specialist help may be necessary. However, fixing instructions and a template are usually supplied with the flange.

If the cylinder is an indirect type it will contain a heat-exchange coil; you must check that this will not obstruct the immersion heater.

The water supply must be turned off and the cylinder drained before the flange is fitted. The cylinder cannot be drained by merely running taps; most modern cylinders include a drain point at the bottom, but if not, drill a small hole in the top at the position where the heater is to be fitted and syphon out the water with plastic tubing.

The immersion heater should be screwed into the flange, using a suitable compound or tape to prevent leakage. Flanges are standard size and will fit all normal immersion heaters. Finally the heater should be tightened using an adjustable or ring spanner or wrench.

The cylinder must, of course, be refilled before the immersion heater is switched on.

How is a cooker connected?

As stated earlier, a cooker of more than 3 kW loading should be on a separate circuit from the consumer unit.

The electrical load and cable size should be determined in accordance with the recommendations in Chapter 9 but a 30 A or 45 A circuit is usually required. For a 30 A circuit the cable should be 4 mm^2 twin and earth PVC-sheathed if fed via an MCB or cartridge fuse but 6 mm^2 if the fuse is a rewirable type. However, it is quite common to use 6 mm^2 cable even when fed via an MCB or cartridge fuse, and this is recommended. For a 45 A circuit the cable sizes are 6 mm^2 and 10 mm^2 respectively. These are minimum sizes and may need to be increased to meet voltage drop requirements – see Chapter 9.

At the cooker, the cable should terminate on either a cooker control unit, which includes a socket outlet, or a simple 30 A or 45 A DP switch (Fig. 41); these items should preferably include pilot lights. A simple switch, without a socket outlet, is preferable since the inclusion of a socket can lead to danger due to flexible cords trailing over the top of the cooker.

The control unit or switch should be fitted at the side of the cooker, rather than behind or above it, at a height of about 1350 mm (minimum) above floor level and within 2 m of the cooker. It should preferably be located well away from the cooker, subject to the 2 m maximum, so that it can be reached without danger in case of a cooker fire.

The final connection to the cooker may be the same size and type of cable as used from the consumer unit to the cooker. This is the only occasion where cable rather than flex is used to link a free-standing appliance to the fixed wiring.

If the control switch is surface-mounted, the cable to the cooker can emerge from the bottom and pass direct to the entry point on the cooker. With a flush-mounted switch, the cable to the cooker should be concealed in the wall (preferably in conduit) down to a special flush connection unit (Fig. 41) fitted at low level, from which the cable passes to the cooker.

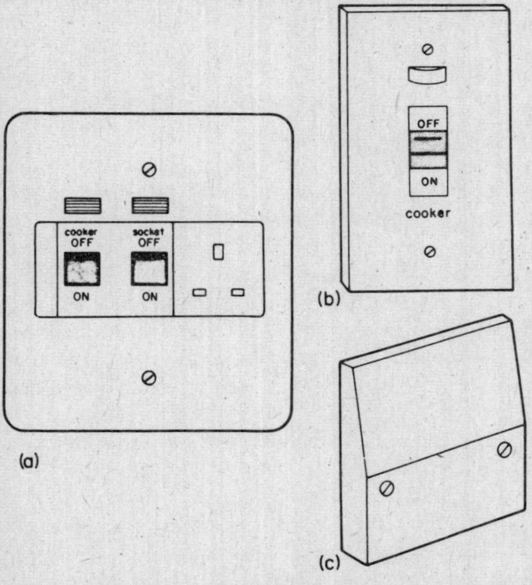

Fig. 41 Cooker control units. (a) with 13 A socket outlet, (b) 45 A double-pole switch, (c) flush connector unit (MK Electric Ltd.)

The connection unit shown in Fig. 41 is available with or without terminals. With the former type the cable is broken and terminated at the unit and the cooker can be disconnected and reconnected very easily, if required. If the type without terminals is used, the cable is unbroken from the control switch to the cooker and is clamped in the connection unit. In either case some slack should be left so that the cooker can be moved for cleaning or maintenance.

When connecting a split-level cooker, the separate oven and hob units may be controlled by the same switch or cooker control unit, provided each one is within 2 m of the control switch. The same cable size must be used throughout from the consumer unit to the control switch and to each section of the cooker. Depending on the layout, a separate cable can be run from the

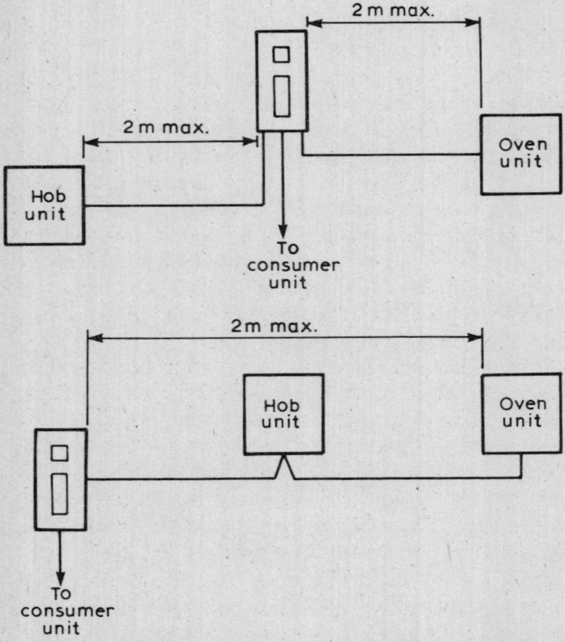

Fig. 42 *Alternative methods of connecting the oven and hob units of a split-level cooker. In each case the same size cable must be used throughout*

control switch to each cooker section or the cable to one section may be looped from the other (Fig. 42).

If the oven and hob units are not both within 2 m of the control switch, a separate control switch must be provided for each section but they may be looped to the same cable from the consumer unit.

A cooker should preferably not be sited adjacent to a refrigerator and the control unit should be at least 2 m from a sink if possible, particularly if it includes a socket outlet.

How is a shower connected?

The electrical loading for a shower may be up to 8 kW and for this load a 45 A circuit is required; this will cater for up to 10.8 kW. For loadings up to 7.2 kW a 30 A circuit is satisfactory.

For a 30 A circuit the cable should be 4 mm^2 flat, twin and earth, PVC-sheathed if fed via a cartridge fuse or MCB; but 6 mm^2 if fed by a rewirable fuse. A 45 A circuit requires 6 mm^2 or 10 mm^2 cable respectively. If there is no spare way on the consumer unit and an extra unit has to be provided, it is advisable to choose one fitted with an MCB or *cartridge* fuse so as to permit the smaller cables to be used; apart from cost, the larger cables are difficult to handle and terminate.

The shower should be on a separate circuit from the consumer unit, fed via a fuse or MCB of the appropriate rating as indicated above. An isolating switch must be fitted within reach of the shower and, since a wall-mounted switch is not permissible (see Chapter 2, page 30), a cord-operated ceiling-mounted switch must be used, preferably with an indicating light which glows when the switch is on. The switch rating should be 30 A or 45 A as necessary. A 45 A switch of this type is available, with indicating light – see Fig. 32, page 59.

Special precautions applicable in a room containing a fixed bath or shower are described in Chapter 2. Care should be taken to comply with these when a shower cubicle is installed in an existing bedroom, with particular reference to any existing socket outlets.

How is a fan connected?

As indicated earlier (page 73), domestic fans may be connected on ring or radial circuits feeding 13 A sockets or on lighting circuits fused at not more than 5 A.

In the former case, the circuit must be fused and should be fed via a fused connection unit fused at 3 A; this may also serve as the control switch, in which case it should be fitted in a convenient position at approximately 1350 mm above floor level and should preferably include a pilot light. If it is more convenient to fit the fused connection unit remote from the fan and use smaller (1 mm^2) cable to the fan, a separate wall or ceiling-mounted switch should be fitted near the fan unless the fan has its own cord-operated switch.

In some cases the fixed wiring can be taken into the fan; if not, e.g. when the fan is fitted in a window pane, the final connection can be made via a flex outlet plate and a short length of flex. Alternatively, a 2 A or 5 A socket outlet or a clock connector (see below) may be used in lieu of the flex outlet plate; do *not* use a 13 A socket.

The fan may be fitted into a wall or window pane and the wiring from the fused connection unit or lighting circuit should be 1 mm^2 flat, twin and earth, PVC-sheathed.

When is a time-delay fan used?

In toilets which have no opening windows or other natural ventilation it is usually compulsory to provide an extractor fan which starts when the room light is switched on and continues to run for a specified period after the room light is switched off. A delay unit is built into the fan.

The wiring is generally as described above for a normal extractor fan, but note that the delay unit requires a permanent supply, connected to the phase conductor, in addition to the wire from the lighting switch.

A wiring diagram is normally supplied with the fan unit, but if not, details should be obtained from the manufacturer.

If the fan unit does not include a means of isolating both supplies, a separate isolator must be fitted adjacent. This can be a DP switch (separate from the lighting switch) and should disconnect the wire from the light switch and the permanent phase supply.

How is a clock connected?

As indicated earlier (page 73), clocks may be connected to either lighting circuits or to circuits feeding 13 A socket outlets, the former usually being more convenient.

The arrangements are generally as for extractor fans as described above except that no control switches are required. The connection to the clock is made via a special clock connector which includes a 2 A fuse and flex outlet. When possible the clock connector and flex should be concealed behind the clock but this may not be possible with small clocks of, say, 150 mm diameter.

What are shaver outlets?

Shaver outlets are of two types as shown in Fig. 43. (a) Shaver supply units suitable for use in bathrooms and similar situations

(a) (b)

Fig. 43 *(a) shaver supply unit, (b) shaver socket outlet (MK Electric Ltd.)*

and (b) shaver socket outlets for use in locations other than bathrooms etc.

Shaver supply unit. This consists of a double-wound isolating transformer complying with BS 3052 or 3535 which provides an earth-free supply at mains voltage. Outputs of 240 V and 115 V are usually provided together with a self-resetting thermal overload cut-out to prevent loads other than shavers being connected. On some models the transformer primary circuit is energized *only* when a shaver plug is inserted.

This shaver unit is suitable for use in washrooms or rooms containing a fixed bath or shower.

Lighting units are produced which include a shaver unit of this type and these may be fitted in bathrooms etc. However, on some lighting units the shaver outlet does not comply with the above, since no transformer is included, and these must not be used in bathrooms etc.

Shaver socket outlet. This is a cheaper version which must not be used in bathrooms etc. but may be used in other locations. It does not include a transformer but incorporates a self-resetting overload device which limits the load to about 200 mA. A 1 A fuse is also included.

Lighting units which include a shaver outlet of this type are produced but these must not be fitted in bathrooms etc.

How are shaver outlets connected?

Shaver outlets of either type may be connected to either 5 A lighting circuits or to circuits feeding 13 A socket outlets; the former method is usually more convenient.

When connected to a circuit feeding socket outlets, the shaver unit should be fed via a fused connection unit fused at 3 A. No further fusing is necessary when connecting to a 5 A lighting circuit.

The final connection to the shaver unit should be by permanent wiring, not by flex.

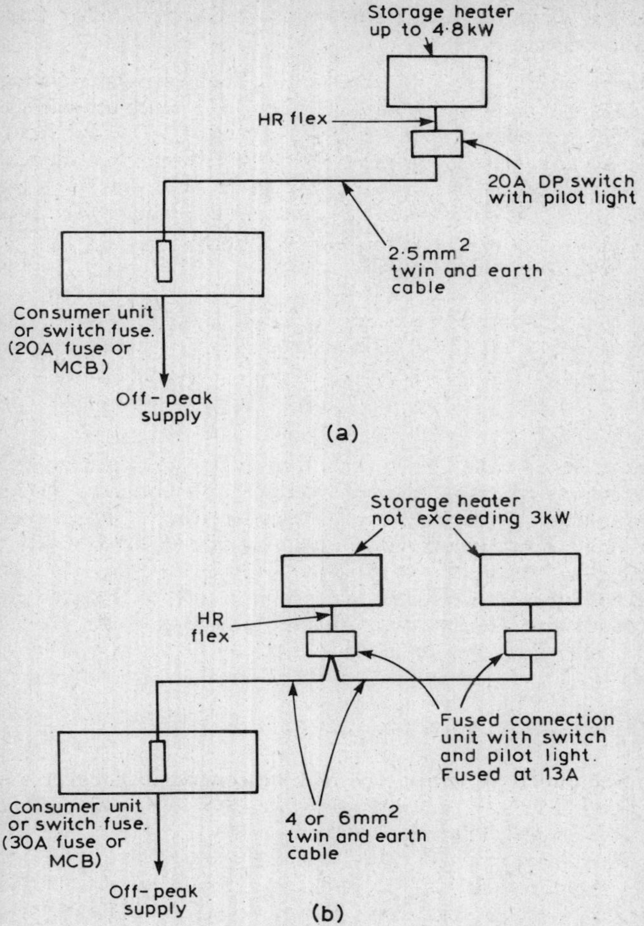

Fig. 44 *Connection of storage heaters: (a) one heater per circuit, (b) two heaters on one circuit*

How are storage heaters connected?

The storage heaters in a dwelling are normally all in use simultaneously under the control of a timeswitch, and it is usual to connect them on individual circuits from a separate consumer unit which is only energized during the off-peak period.

The electrical load per heater may be up to 4.8 kW (20 A) and it is advisable to provide 20 A circuits even if the initial loads are less. This will permit the installation of larger heaters – up to 4.8 kW – at a later date, without rewiring. The cable should be 2.5 mm^2 flat, twin and earth, PVC-sheathed, terminating on a 20 A DP switch (unfused), with flex outlet and preferably with pilot light, fitted adjacent to the heater. The final connection to the heater should be in three-core heat-resisting flex having a current rating at least equal to that of the fuse feeding it.

However, two heaters, each not exceeding 3 kW, may be connected on a common 30 A circuit; a fused connection unit, fused at 13 A, with DP switch and flex outlet, and preferably a pilot light, should be fitted adjacent to each heater. The 3 kW maximum is necessary to prevent possible overloading of the fused connection unit. In this case the cable size should be 4 mm^2 if fed via a cartridge fuse or MCB but 6 mm^2 if fed by a rewirable fuse (Fig. 44).

When storage heaters are fan-assisted it is necessary to provide a permanent supply to the fan in addition to the off-peak supply to the heater. Special switch units are available to provide for the

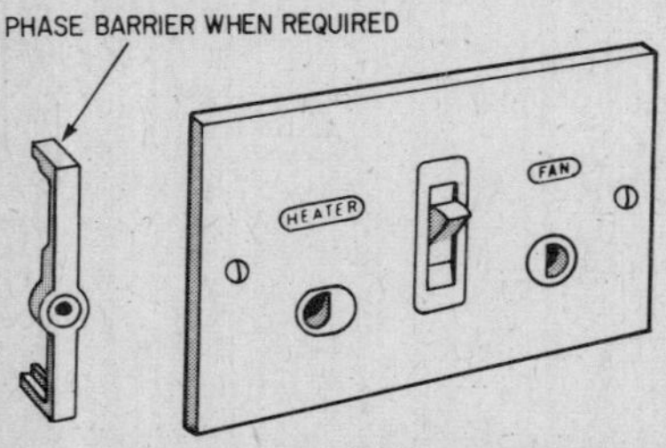

Fig. 45 *Switch for fan-assisted storage radiator with 25 A double-pole switch controlling both supplies (MK Electric Ltd.)*

connection of the two supplies; these units include a 25 A DP switch, which controls *both* incoming supplies, and two flex outlets. Fig. 45 shows a typical unit. The supply to the fan may be taken from a convenient lighting circuit without further fusing, or from a circuit feeding 13 A socket outlets via a fused connection unit fused at 3 A. The cable to the switch from the lighting circuit or fused connection unit can be 1 mm^2 flat, twin and earth, PVC-sheathed.

In some cases the fans are remotely controlled by a thermostat and/or timeswitch and terminals are usually included on the switch units for these items.

How are the controls for a central heating system connected?

Usually, unless specified otherwise, the pump and controls for a central heating system are connected to a ring or radial circuit feeding socket outlets. It is, however, preferable to connect them to a separate 5 A circuit from the consumer unit. This prevents the central heating being put out of action by a fault on another circuit, which could have serious consequences in frosty weather if the dwelling is unoccupied. A separate circuit also permits other circuits to be disconnected whilst leaving the heating still operative, e.g. if the house is left empty.

For similar reasons, it is undesirable to connect the central heating via an ELCB which also feeds other items.

Equipment, such as gas boilers, which may need to be disconnected or removed for maintenance should preferably be connected via a plug and socket outlet rather than fixed wiring; they can then be disconnected easily by non-electrical personnel. The appropriate authority, e.g. the gas board, should be consulted if necessary. The plug and socket should preferably *not* be the standard 13 A type.

What are timeswitches used for?

They are used to switch appliances on and off automatically at pre-determined times. Typical cases are off-peak loads, fans on storage heaters, immersion heaters, central heating controls, and security lighting.

Timeswitches normally operate at mains voltage by a small synchronous motor and for domestic work are usually single pole. The make/break contact must always be connected in the phase side of the circuit. The rating marked on the timeswitch, e.g. 20 A, is the maximum current the switch will carry and this value must not be exceeded. 20 A is normally adequate for domestic use.

Timeswitches are available with 'spring reserve' and these will continue to run for a specified period, e.g. 24 hours, if the mains supply is interrupted; these are more expensive than the standard type and are not normally justified on domestic work.

When a timeswitch is used to control an appliance which operates on the off-peak supply it requires a *permanent* feed to the motor in addition to the off-peak supply, since the timeswitch must run continuously. In such a case, make sure that you buy the correct type since some versions are suitable for one supply only, which feeds the motor and the load. The timeswitch motor requires only a very small current and may be connected to a 5 A lighting circuit or, via a fused connection unit, to a circuit feeding socket outlets. Make sure that the mechanism of the timeswitch is connected to the protective conductor.

It is often convenient to fit an override (or bypass) switch so that the timeswitch can be bypassed when desired; this is connected across the two output terminals of the timeswitch. If the bypass switch is DP, the second pole can be wired to an indicating lamp to show when the switch is operated. Some timeswitches include a push-button type bypass switch (otherwise called an 'anticipatory' switch).

Timeswitches are also available which plug into a standard 13 A socket outlet and into which a 13 A plug can be inserted. These can be used to control portable appliances e.g. cassette-radios; they should not, however, be used to control an immersion heater – see page 77.

A timeswitch must *never* be used to control a radiant fire since this can constitute a fire risk if the fire is left near a flammable object.

The use of timeswitches to control immersion heaters is described on page 78.

How is a heat/light unit connected?

A heat/light unit consists of a circular, ceiling-mounted heater rated at about 750 W, with a filament lamp in the centre.

These units must *not* be connected to a lighting circuit but should be fed from a ring or radial circuit feeding 13 A socket outlets. The connection unit should be fused at 10 A or 13 A and the cable from this to the heat/light unit should be 1.5 mm^2 flat, twin and earth, PVC-sheathed. See Fig. 46.

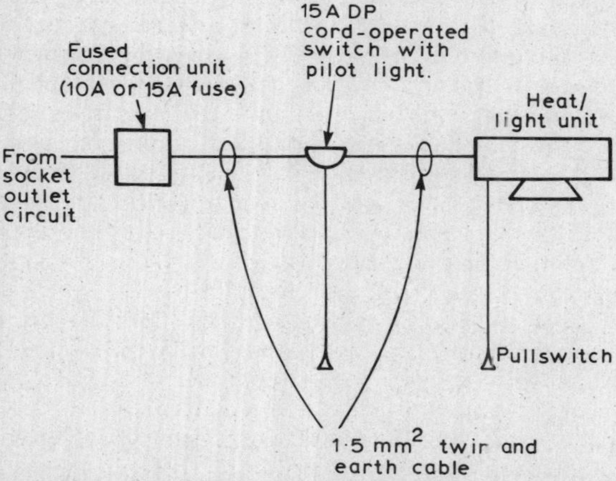

Fig. 46 *Typical wiring for heat/light unit fitted in bathroom*

If the heater is being fitted in a bathroom the fused connection unit must be fitted outside the room and a separate DP cord-operated, ceiling switch, with pilot light, fitted inside; this is additional to any cord-operated switches on the heat/light unit. The fused connection unit may be fitted at any suitable position; it is often convenient to fit it adjacent to a socket outlet.

In a room other than a shower- or bathroom, the fused connection unit may be in the same room if desired, in which case it can serve as the isolating switch and should include a DP switch and pilot light. If the connection unit is not in the same room a separate wall or ceiling-mounted DP switch with pilot light should be fitted within the room.

Switching for the heating and lighting elements on the heat/light unit is usually provided by one or more cord-operated switches on the unit.

If a heat/light unit replaces a normal luminaire it must *not* be connected to the existing wiring. This wiring should be disconnected and either removed, or made safe by suitable connectors inside a junction box.

Can I convert existing 15 A round-pin socket outlets to 13 A type?

Yes, *provided* that (a) the existing sockets are wired in at least 2.5 mm^2 or 7/.029 cable and the cable is in satisfactory condition (any 7-strand conductor may be assumed to be at least 7/.029) and (b) there is not more than one socket outlet on a circuit fused at 15 A or more than two on a 20 A circuit, or a circuit on which the fuse can be uprated to 20 A.

If there are more than two sockets on a circuit they should not be converted until the circuit has been rewired. The existing circuit arrangements can readily be ascertained by removing fuses in turn and checking which sockets are affected.

If the existing 15 A socket outlets do not have a protective (earth) conductor, a separate 2.5 mm^2 PVC-insulated green-yellow single-core cable should be run from the consumer unit to the socket; this may be looped between socket outlets on the same circuit.

Socket outlets of lower ratings than 15 A should not be converted since the wiring will be inadequate for loads of up to 13 A.

6

Outbuildings

How is the supply to an outbuilding provided?

Typical cases involving outbuildings are garages, sheds, greenhouses etc.

The supply should preferably be taken from the intake position in the house, using a spare way on the consumer unit if available. If not, a separate switchfuse or additional consumer unit should

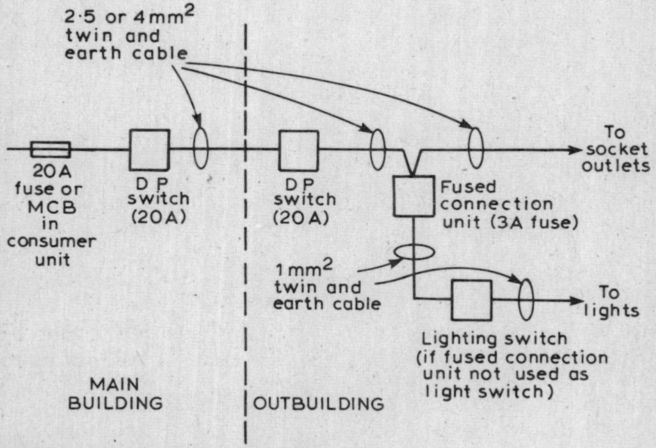

Fig. 47 *Typical supply to outbuilding*

be fitted. If a way on a consumer unit is used it is preferable to fit a separate DP isolating switch, suitably labelled, adjacent to the consumer unit and controlling the cable to the outbuilding. Normally a 20 A or 30 A circuit will suffice, depending on the estimated load.

In the outbuilding the cable should terminate on a consumer unit, or a splitter (see below), or a DP isolating switch, depending on the requirements in the outbuilding. (A splitter is a switchfuse which provides two outgoing fused circuits.) This should preferably be fitted near the entrance at not more than 1500 mm above floor level. If the installation consists only of a few 13 A socket outlets and lights, a 20 A circuit and isolating switch may suffice and the sockets can be fed direct from this. The lights may then be fed via a fused connection unit, fused at 3 A, connected to the socket outlet circuit. See Fig. 47. The fused connection unit can also serve as the lighting switch if desired, in which case it should be fitted near the entrance at about 1350 mm above floor level.

If the requirements are more extensive, a small consumer unit or splitter may be necessary.

What type of cable should be used?

The cable between the buildings may be (a) flat twin and earth PVC-sheathed as described in Chapter 3, (b) mineral-insulated, copper-sheathed (MICS) with PVC sheath overall, or (c) PVC-insulated PVC-sheathed, armoured cable. Flexible cable should *not* be used.

MICS and PVC-sheathed armoured cables involve special termination arrangements which may not be within the scope of unskilled persons and they are also more expensive than the flat twin PVC-sheathed type. They are, however, very suitable for laying direct in the ground and need not be in conduit or pipes. If PVC-sheathed unarmoured cables are buried in the ground they *must* be enclosed throughout in galvanized metal conduit or high-impact PVC conduit or in earthenware or steel pipes. The copper sheath of MICS cable or the wire armour of PVC armoured cable can be used as the protective conductor but in the latter case it is preferable to use a *three-core* cable with one core being used as the protective conductor in parallel with the armour wires. If a two-core armoured cable is used, which means that the armour is the sole means of earthing, the circuit should preferably have earth-leakage protection within the house. Earth-leakage protection is described in Chapter 2.

If MICS or PVC-sheathed armoured cable is used for the underground cabling it will probably be cheaper to use this expensive cable for the underground section only and PVC-sheathed unarmoured cable inside the building at each end; the connections should be made at metal junction boxes with fixed screwed terminals fitted inside the buildings.

An advantage of PVC-sheathed twin and earth cable is that a separate continuous protective conductor is provided throughout the installation.

How is the cable size determined?

For a 20 A circuit, a 2.5 mm^2 flat PVC-sheathed, twin and earth cable may be used whilst a 30 A circuit requires 4 mm^2 cable if protected by a cartridge fuse or an MCB or 6 mm^2 if the fuse is a rewirable type. However, for a 20 A circuit it may be well worthwhile to use the 30 A cable size to allow for possible future loads since the extra cost is very small and the provision of a larger cable at a later date would be expensive.

The above cable sizes are subject to voltage-drop requirements and if a long cable run is involved it must be checked that the volt drop does not exceed the permissible limit, i.e. 2.5 per cent, which is 6 V on a 240 V supply (see Chapter 1).

The method of calculating voltage drop is described in Chapter 9, and if the above cable sizes give an excessive drop, the size must be increased.

As an example, if we assume a maximum estimated load of 25 A and a route length of 50 m the voltage drop will be 13.8 V for a 4 mm^2 cable, 8.9 V for 6 mm^2 and 5.25 V for 10 mm^2. The 10 mm^2 cable must therefore be used; this will be satisfactory since the voltage drop within the outbuilding should be negligible and the current rating is well above that of the fuse or MCB, which will be 30 A.

How is the cable installed?

The cable may be run above or below ground depending on circumstances.

If the outbuilding is connected to the house by a wall or by a *substantial* solid fence the cable may be run on this. If flat PVC-sheathed or MICS cable is used, any cable at less than 1500 mm above ground level should be enclosed in galvanized metal or high-impact PVC conduit with bushes at each end. PVC armoured cable will not normally need mechanical protection except where there is abnormal risk of damage. Metal conduits running above ground for the protection of cables should be earthed, using 2.5 mm^2 PVC-sheathed single-core cable connected to the conduit by a suitable clamp. The other end of this cable should be connected to earth at either the intake position in the house or the entry point in the outbuilding.

Cable routes should be carefully chosen to reduce the risk of damage, e.g. cables should be run in corners and at high level. They should be run inside the buildings as far as possible, i.e. in floor or ceiling voids, so as to keep the external cabling to a minimum.

If there is no suitable wall or fence available and flat PVC-sheathed cable is run overhead, no support is strictly necessary for spans of up to 3 m provided strain is avoided, but it is still preferable to support the cable on a wood batten or by other means or to enclose it in conduit. For longer spans the cable must be supported by the above means or by a catenary wire – see below.

Overhead cables should be at least 3.5 m above ground level provided the situation is not accessible to vehicular traffic; otherwise the clearance should be at least be 5.2 m.

However, if a long run is involved, overhead wiring may be unsightly and subject to damage or deterioration, and it may be preferable to bury the cable. This gives a more secure, permanent and unobtrusive installation.

How is an underground cable installed?

Cables should be buried at a minimum depth of 450 mm, measured to the top of the cable but under cultivated areas the depth should be at least 600 mm.

MICS or PVC-sheathed armoured cables should be protected by earthenware cover tiles or other suitable means. Flat PVC-sheathed unarmoured cables may be laid underground but they *must* be enclosed in pipes or ducts as described above.

Cables laid direct in the ground, i.e. not in pipes, should be laid on a layer of about 50 mm of sand or sifted soil with a similar layer above the cable. No sharp stones or flints should be near the cable as they could damage the PVC or copper sheathing.

Cable routes should be carefully chosen; cultivated areas and concrete or paved areas should be avoided when possible and crossing of lawns should also be avoided due to the difficulty in getting satisfactory reinstatement. Runs along the edges of borders or lawns are least likely to be disturbed.

What is catenary wiring?

This system consists of a single or stranded galvanized steel wire stretched between the buildings or other supports and to which the cable is attached. The wire is attached to eye bolts, one of which includes a tensioning device. The eye bolts should be at least 300 mm higher than the minimum clearance required above the ground (see above) so as to allow for sag. For long runs, intermediate supports may be necessary.

The cable may be any of the three types mentioned earlier and is attached to the catenary wire by clips or slings. If slings are used, the cable is some distance below the catenary wire, and this should be borne in mind when fixing the eye bolts so as to obtain the minimum height required.

What about earth-leakage protection?

As indicated in Chapter 1, any socket outlet which may be used to feed portable appliances outdoors via flexes *must* have earth-leakage protection operating at not more than 30 mA and this requirement often applies to sockets in outbuildings. However, it is strongly advisable to provide such protection, even when not compulsory, if the conditions are likely to be damp or the floor is concrete, and portable tools, inspection lamps or other appliances may be used, e.g. a typical garage.

The types of earth-leakage circuit breaker available are described in Chapter 2 and the device may be located in either the house or the outbuilding. The former is preferable since the supply cable to the outbuilding is protected against earth leakage as well as the installation within the outbuilding. If the earth-leakage protection is fitted in the outbuilding, a consumer unit incorporating a residual current device (RCD) in lieu of a main switch may be used; alternatively, if only a single or twin socket-outlet is involved, a combined socket outlet/RCD unit (see Fig. 15) may be used. The other methods described in Chapter 2 are also available.

RCDs in the above circumstances should operate at not more than 30 mA.

In greenhouses it is undesirable to provide earth-leakage protection on circuits feeding heating appliances since, if it operates, growing crops may be ruined.

What materials should be used in outbuildings?

The cable can be normal flat twin and earth PVC-sheathed cable as used for house wiring. When running along bare brick walls it may be easier to fix the cables on wood battens rather than direct on the wall. Mechanical protection should be provided when necessary, e.g. at low level.

Accessories may be of the normal moulded plastic type but if there is undue risk of damage, the metal-clad type should be used. If the conditions are damp or wet (e.g. in a greenhouse), special watertight, sealed accessories should be used. Socket outlets or fused connection units in greenhouses should preferably include pilot lights.

7

Bells and intruder alarms

How do bells work?

Bells on domestic premises normally work at extra-low voltage (ELV), usually about 5–8 V, which is obtained from dry batteries or a mains transformer. Bells are usually of the 'trembler' type (see below) which operate on either DC (batteries) or AC (mains transformer). It is, of course, essential that the power supply matches the bell voltage; a normal dry cell has a voltage of 1.5 V so a 6 V bell requires at least four cells, possibly more if long wiring runs are involved.

The primary side of the bell transformer is connected to the normal mains supply and typical outputs on the secondary side are 3, 5, or 8 V according to the terminals connected – see Fig. 48.

The mains wiring to the transformer must be in normal mains voltage, 1 mm^2, twin and earth, PVC-sheathed cable which may be connected direct to a 5 A lighting circuit, or to a circuit feeding 13 A socket outlets, (via a fused connection unit, fused at 3 A) or to a separate 5 A way on the consumer unit. The mains supply to the transformer should be connected via a 5 A switch fitted adjacent.

The ELV wiring to the bells and pushes may be lightly-insulated, twisted twin bell wire (usually 0.4 mm^2) running along walls etc. and fixed with insulated staples or insulated, flat-headed pins fixed between the two wires. Alternatively, a flat cable in strip form with adhesive back may be used. This bell

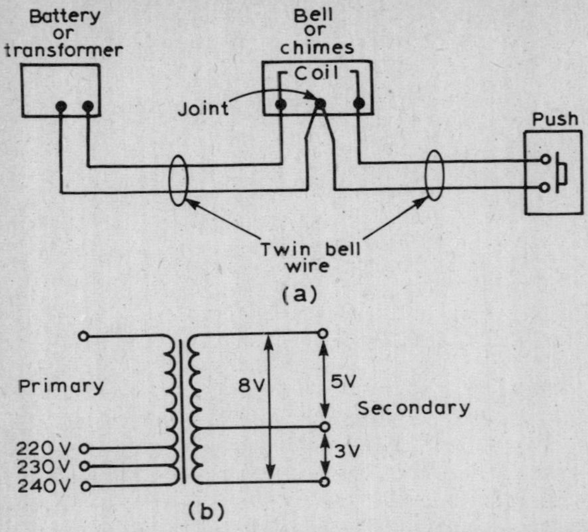

Fig. 48 (a) Connections for typical system, (b) Typical transformer with 3 V, 5 V or 8 V output

wiring must be quite separate from wiring carrying mains voltage. Fig. 48 shows a typical system.

If an illuminated push is used, it is essential to supply the system by a mains transformer, not batteries, since the lamp in the push uses energy continuously and batteries would be run down very quickly.

Batteries should preferably be fitted near the bell. The transformer is often fitted adjacent to the consumer unit but if this involves a long run (above about 15–20 m) of ELV wiring, it is best to locate it as near as possible to the bell.

How does a trembler bell work?

A trembler bell (Fig. 49) includes a make-and-break contact in series with the coil circuit. When the push is first pressed, the

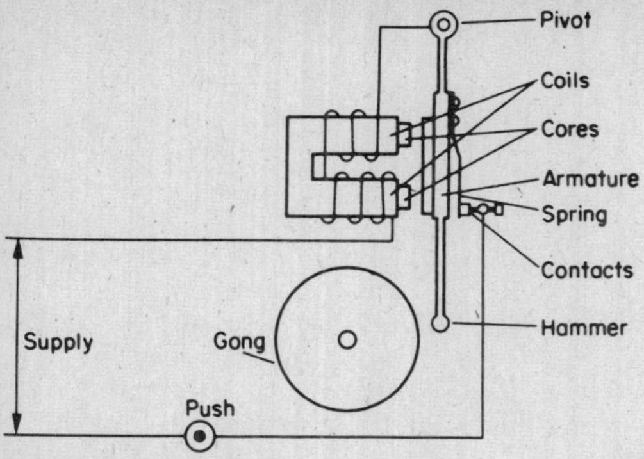

Fig. 49 *Trembler bell*

current through the coils energizes the electro-magnet and the armature is attracted, causing the hammer to strike the gong. At the same time, the contact is broken, the coils are de-energized and the armature returns to its original position under the influence of the spring. The contact is now re-made and the process is repeated. This backwards and forwards 'trembler' movement of the armature continues as long as the push is pressed, causing the bell to ring.

When is a buzzer used?

A buzzer operates on the same principle as a trembler bell but it has no gong and produces a buzzing sound instead of a ring. A buzzer can be used when a loud signal is not required; or, if two separate systems are in use, one system may use a bell and the other a buzzer so as to indicate the source of a call.

How do chimes work?

Door chimes provide a distinctive two-tone signal by means of two tone bars (or tubes) and a spring-loaded plunger operated by a solenoid. When the door push is pressed, the solenoid operates and the plunger strikes one of the tone bars. When the push is released, the plunger springs back and strikes the other tone bar, thus producing the ding-dong signal. For the best effect the chimes should be fitted about 2 m above floor level.

Chimes are also available which give distinctive signals when there are two door pushes at different locations; e.g. the push at one door may cause the normal double chime whilst the push at the other door gives a single note, thus indicating the source of the call.

Chimes giving other signals, e.g. a warbling note or even a well-known tune, are also available.

Chimes may be operated from batteries or a mains transformer as indicated above and the voltage required is usually 6–8 V, or more if long (above 15–20 m) wiring runs are involved. The operating voltage is stated on the equipment and should be carefully checked before buying the transformer or batteries. If in doubt, consult the manufacturer.

Some manufacturers supply complete 'do-it-yourself' kits containing all the items necessary for a bell or chimes installation, including bell wire, staples, screws etc., but not normally batteries or transformer.

What are the requirements for a bell transformer?

The transformer should comply with BS 3535. The primary and secondary windings must be entirely separate from each other and the secondary winding must be insulated from earth (Fig. 48), i.e. it must be a safety-isolating transformer specifically designed for supplying bells, buzzers or chimes. Auto transformers must not be used.

Bell transformers are usually designed for 220/240 V on the primary side and, in the UK, the 240 V tapping should normally be used. An output of 1 A is normally adequate for one bell or chime.

The power consumption of a bell transformer may be ignored in assessing the total load on an installation.

What is an intruder alarm?

An intruder alarm system is intended to give an alarm if unauthorized entry is made to the premises whilst the alarm system is operative. For domestic premises the alarm signal usually consists of a loud-ringing bell located outside the building.

The alarm may be actuated by door or window contacts, pressure-pad contacts beneath carpets, or invisible rays. The system can be switched off whilst the house is occupied but should be switched on again during the night when the occupants are in bed.

Systems are available in kit form, complete with installation instructions, and it is preferable, and usually cheaper, to use one of these rather than to buy the components separately.

How is the power supply obtained?

It is undesirable that an intruder alarm system should be fed from the mains supply *only* since if this supply fails, or is disconnected by a potential intruder, the system will not operate. The system should therefore preferably be fed via batteries as well as from the mains so that it still operates if the mains supply is interrupted. Batteries are usually of the secondary rechargeable type but the standard disposable dry type is used in some cases.

The voltage supplied to the alarm-operating devices *must* be ELV, i.e. not exceeding 50 V, for safety reasons. When this is obtained from the mains it *must* be via a safety isolating transformer, as described earlier. The operating voltage for intruder alarm systems is usually 12 V DC, obtained via a step-down transformer and rectifier.

If the door or window contacts were connected directly to the mains supply and this caused death or injury (even to an intruder), the householder could be held responsible.

All mains wiring must, of course, be carried out in normal mains-voltage cable.

The ELV wiring to the alarm-operating devices, i.e. door or window contacts, etc., may be in lightly-insulated bell wire. This should be concealed as far as possible, particularly at doors and windows etc. for obvious reasons, and should preferably be enclosed in metal conduit or channelling at vulnerable positions.

8

Inspection of existing installations

How can I tell if rewiring is necessary?

This is a very common problem and it can be difficult to get unbiased advice. Your local electricity board or a registered electrical contractor will usually carry out an inspection and provide a report on the state of the installation and the action recommended, together with a quotation if applicable. They will usually do this without charge, but remember that they are looking for rewiring work.

However, it is quite possible for a reasonably capable person to form an opinion on the state of the wiring and whether rewiring is necessary.

Age. If the wiring is more than 30 years old you may take it that rewiring is necessary. Between 20 and 30 years is a borderline case but rewiring will almost certainly be required within the next few years.

Type of cable. Lead sheathed cables with rubber-insulated conductors were widely used in the past but have not been used now for many years. Any such cables should *definitely* be replaced.

Cables with rubber insulation and sheathing (commonly called 'TRS' or 'VIR' cables) were also commonly used but have been superseded by the PVC-insulated type. These cables can be assumed to have had their useful life and should be replaced. This also applies to single-core, rubber-insulated cables in conduit.

(You can easily distinguish rubber-insulated cables from PVC-insulated by the fact that on the former type the wires are tinned whereas with PVC insulation the conductor is not tinned and has a natural copper colour.)

Whilst the above cables may be working satisfactorily so long as they are not disturbed, it will usually be found that the insulation on the conductors is brittle at outlet points and breaks when disturbed.

Cables should be examined along the runs, i.e. below floorboards and in roof spaces, as well as at terminations. They may have been renewed for only a short length near the ends; this situation can arise quite legitimately, if the old main switchgear is replaced by a modern consumer unit and the existing cables are too short.

Conduit installations. It was quite common in the past for houses to be wired in lightweight, grip-type conduit using single-core, rubber insulated taped and braided cables. The cables will need replacing as indicated above but, in addition, such conduits do not provide satisfactory earth continuity and also they frequently terminate short of the outlet points, such as ceiling roses or socket outlets, leaving the last few mm of wiring exposed. See Fig. 17. This situation is unsatisfactory and rewiring should be carried out using flat, PVC-insulated, PVC-sheathed cables with earth conductor. You can run these in the existing conduits, when convenient, but the conduit should not be used as the protective conductor. See also Chapter 9.

Main switchgear. It has been customary for many years to use consumer units (see Chapter 2) at the intake position to accommodate the consumer's main switch and fuses or MCBs. On old installations it may well be found that the switchgear consists of an assortment of switchfuses, fuseboards, isolators and other items. However, such switchgear does not necessarily indicate that rewiring is necessary (consumer units are not compulsory) but it does show that the installation is probably quite old.

If the switchgear is in poor condition it should be replaced by one or more modern consumer units. If it is in reasonable condition it can be retained but if you *do* carry out rewiring you should always fit consumer units if they do not already exist. Conversely, the presence of consumer units does not necessarily imply that the wiring is satisfactory since the switchgear may well have been replaced without renewing the wiring; this situation is quite common.

It is still possible to find installations with DP fusing i.e. with fuses in both the phase and neutral conductors; this should be changed to SP fusing *immediately*.

Accessories. The type and condition of items such as lighting switches, socket outlets, etc. will give you an indication of the age and state of the wiring. If the switches are the old round type, mounted on wooden blocks, they will obviously have been in for many years and are due for replacement. Socket outlets with round pins, and other non-standard types, are now obsolete but are still in fairly common use and should not necessarily be condemned; they may have been installed to match an existing installation. However, if the accessories are old and worn, with arcing at switches and the plugs loose in the sockets, you should replace them by modern types.

The presence of modern accessories, such as 13 A socket outlets, does not necessarily mean that the wiring is satisfactory since it is quite common for new fittings to be connected to the existing wiring. The wiring should be inspected by removing switches, ceiling roses, socket outlets etc. as necessary.

If rewiring *is* carried out *all* accessories should be replaced unless they are already of a modern type and in good condition.

Insulation resistance. If possible, you should measure this – you may need to get expert assistance. Low readings indicate that action is required but satisfactory readings do not necessarily mean that the wiring is in good condition. Even very old wiring may have good insulation values, particularly if it is dry and enclosed in non-earthed conduit or is not adjacent to other earthed conductors.

Supplies to outbuildings. These should be given special attention since it is quite common to find unsatisfactory or even dangerous conditions.

Flex should *never* be used for a supply to an outbuilding; the cable should be of the type specified in Chapter 6 and installed as detailed therein. Cables which run at low level without mechanical protection, or which are strung along hedges or rickety fences, or which are otherwise unsatisfactory, should be replaced.

As indicated in Chapter 6, it is preferable (compulsory in some cases) that sockets in outbuildings should have earth-leakage protection and this should be seriously considered, whether or not rewiring is necessary.

What if the installation does not comply with the latest edition of the IEE Wiring Regulations?

These Regulations are amended from time to time to comply with current developments and ideas. However, the changes are not retrospective and an existing installation cannot be condemned merely because it does not comply with the latest edition, provided it meets the minimum statutory requirements – see Chapter 1.

However, if rewiring *is* carried out, the work should comply in all respects with the current IEE Wiring Regulations.

When a property is being sold, or when a survey and opinion is obtained from a specialist, it is sometimes claimed that rewiring is necessary since the installation does not meet the current IEE Wiring Regulations. This should not necessarily be accepted.

What about earthing?

When the incoming mains supply is by underground cable it is usual for the electricity board to provide an earthing terminal at the intake position, connected to the sheath and/or armour of the supply cable. On some old installations this may not apply and the only earth connection may be to a water pipe.

It is not now permissible for water or other service pipes to be used as the *sole* means of earthing and if this situation applies you should take steps to remedy the position. First, discuss the matter with the electricity board; if the supply is underground they may well agree to provide an earth terminal but in any case they will make recommendations.

If the supply to the property is by overhead wires, with no incoming earth connection, there may be a fault-voltage-operated ELCB (see Chapter 2); this type has now been superseded by the residual-current-operated type but some are still in use. Whichever type is installed, it should be tested by means of the test button; this is only a partial test but does test the mechanical operation. The residual-current type of ELCB is nowadays preferred to the fault-voltage-operated type and if rewiring *is* carried out in the above circumstances the former type should be fitted, the fault-voltage-operated type being removed.

A quick test that an earth connection exists at socket outlets can be made by a mains voltage test lamp connected between the phase (L) and earth terminals of a plug as indicated in Chapter 10, although this does not necessarily imply that the impedance of the earth connection is satisfactory.

In addition, the earthing connections at the intake position should be examined; check that clamps to pipes or cables are satisfactory and that there is no sign of corrosion.

How much of the installation need I rewire?

It is only necessary to rewire the sections in which you consider the wiring to be unsatisfactory after inspection as indicated above. Most installations have had extensions added over the years and this wiring may be quite satisfactory.

In cases where new socket outlets have been added and the cable is PVC-insulated and PVC-sheathed type in good condition, with earth conductor, this can often be utilized to form part of a new ring system for 13 A sockets – see Chapter 5. Such wiring may have been installed before the metrification of cable sizes and may have 7/.029 conductors; this size is larger than 2.5 mm^2 and you can therefore use it as part of a ring circuit. (A 7-strand cable may be assumed to be at least 7/.029.)

It will usually be found that the lighting wiring is the oldest in the property and in most need of rewiring.

Do I need permission to rewire?

If you own the property you have no need to get anyone's permission to rewire or to carry out any other electrical work in it.

For rented property you must get the landlord's permission. This is not likely to be withheld (unless you are expecting him to pay for it) provided the work is carried out in a responsible manner and does not inconvenience any other tenants or cause damage to the property. No other permission is necessary.

9

Planning the installation and rewiring

What is a final circuit?

A final circuit is one which is connected to a separate fuse or MCB (often called a 'way') on the consumer unit or distribution board. The wiring of each final circuit must be entirely separate from that of any other circuits. This point should be watched

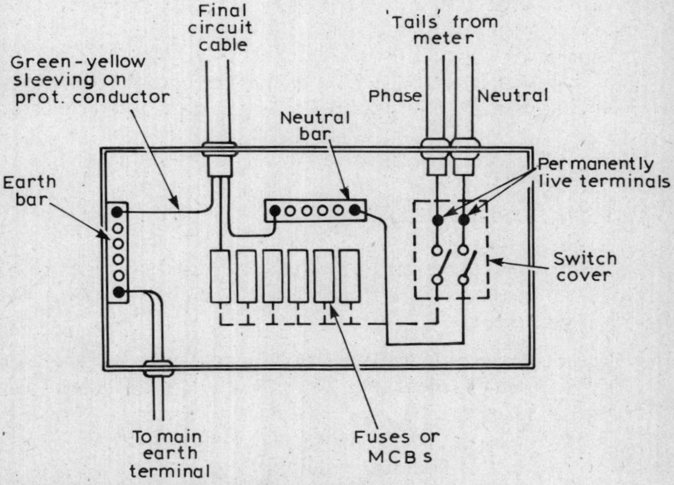

Fig. 50 *Typical connections at consumer unit*

when two-way switching is used on lighting circuits as there is sometimes a temptation to use the phase conductor of one circuit and the neutral from another, which is incorrect.

Fig. 50 shows a typical layout of circuits at a consumer unit.

How are the circuits arranged?

On a normal domestic installation, final circuits are usually required for: lighting (5 A), socket outlet ring circuits (30 A), socket outlet radial circuits (20 A or 30 A), cooker (30 A or 45 A), immersion heater (15 A or 20 A), possibly a freezer (5 A or 15 A), central heating controls (5 A), shower (30 A or 45 A) and possibly an outbuilding (usually 20 A or 30 A).

Some of the above may not be required in all cases and for some circuits, such as lighting and socket outlets, more than one circuit may be necessary.

If storage heaters are being installed, a separate consumer unit is required, connected to the off-peak supply. One way on the consumer unit is required for each heater except in cases where two heaters are supplied by one circuit – see page 88.

It is well worthwhile to label the cables where they leave the consumer unit, using white adhesive tape or other suitable means.

How is the consumer unit chosen?

Consumer units are described in Chapter 2 and the type and size should be based on the information therein, covering the following points:

- Number of ways. Allow one or two spare ways; an 8-, 10- or 12-way is usually suitable on a normal domestic installation. A unit with fewer than eight ways should not normally be fitted.
- Size of main switch. 80 A or 100 A is normally preferable but a 60 A may suffice in a very small dwelling.
- What earth-leakage protection is to be provided, if any.

- Whether MCBs, cartridge fuses or rewirable fuses are to be used.
- Whether the consumer unit is to be flush or surface-mounting and metal or insulated case.

How is the load on a cooker calculated?

The possible load on a large cooker, assuming that all the ovens, hotplates etc. are in use at once, can be up to 16 kW or even more. However, it is highly unlikely that this load will ever occur in practice; even when the various sections are switched on they are controlled by thermostats which disconnect them at intervals.

The method recommended in the IEE Wiring Regulations for calculating the load on a domestic cooker is as follows: (a) the first 10 A plus (b) 30 per cent of the remainder of the rated current plus (c) 5 A if a socket outlet is included on the control unit.

Thus assuming a nominal 12 kW cooker, which would take 50 A at 240 V, the estimated load would be: 10 A, plus 30 per cent of 40 A, plus 5 A for the socket outlet, giving a total of 27 A. Thus a 30 A circuit would suffice but it might be preferable to provide a 45 A circuit to allow for a larger cooker at a later date.

How do I calculate the cable size?

On domestic work the cables sizes for PVC-sheathed cables are usually 1 mm^2 for lighting, 2.5 mm^2 for socket outlet ring circuits, and 4 mm^2, 6 mm^2 or 10 mm^2 for cookers or showers as indicated in Chapter 5. The above sizes are all adequate with regard to current rating, but in some cases when long runs are involved it may be necessary to check the voltage drop and a typical example follows.

Assume a cooker circuit with an estimated load of 38 A, protected by a 45 A fuse or MCB and supplied by a twin and earth, flat, PVC-sheathed, twin and earth cable, installed under normal conditions, route length 50 m.

The current rating of the cable must be at least equal to that of the protective device i.e. 45 A *not* the actual load of 38 A. A 10 mm^2 cable will be satisfactory for current rating but we must check that the voltage drop from the supply point to the cooker will not exceed 2.5 per cent of 240 V, i.e. 6 V (see Chapter 1).

From Table 2 the voltage drop in the above 10 mm^2 cable is 4.2 mV per ampere per metre, so the drop in this case will be 38 × 50 × 4.2 mV which gives 7.98 V which is too high. (Note that the *actual* current, i.e. 38 A, is used in this case.)

We will now consider the next larger cable size which is 16 mm^2 and for which the voltage drop is 2.7 mV/A/m. The

Table 2. Current ratings and voltage drops for twin, PVC-insulated, PVC-sheathed cables with or without protective conductor

Conductor cross-section (mm^2)	*Enclosed in conduit or trunking*		*Clipped direct to a surface or on a cable tray, and unenclosed*	
	Current rating (A)	*Volt drop per ampere per metre (mV)*	*Current rating (A)*	*Volt drop per ampere per metre (mV)*
1.0	14	42	16	42
1.5	18	28	20	28
2.5	24	17	28	17
4	32	11	36	11
6	40	7.1	46	7.1
10	53	4.2	64	4.2
16	70	2.7	85	2.7
25	79	1.8	108	1.8
35	98	1.3	132	1.3

The above current ratings apply when the cables are protected by MCBs or cartridge fuses. If rewirable fuses are used, the ratings should be multiplied by 0.725.

The ratings are also based on an ambient temperature of 30°C and are for cables run singly. For higher temperatures and other conditions of installation the ratings may need to be reduced and the IEE Wiring Regulations, Appendix 9, should be consulted.

voltage drop will now be 5.13 V and since this is within the limit of 6.0 V this cable will be satisfactory for both current rating and voltage drop.

Table 2 gives the current ratings and voltage-drop values for twin and earth PVC-sheathed cables as used for house wiring.

It is important that the permissible value for voltage drop is not exceeded since all electrical equipment suffers a loss of power and efficiency if operated at less than its design voltage.

Table 3 gives the current ratings for flexible cords and the maximum weights they will support.

Table 3. Current ratings and maximum weights supportable for flexible cords

Cross-sectional area (mm^2)	*Current rating (A)*	*Maximum mass supportable by twin flexible cord (kg)*
0.5	3	2
0.75	6	3
1.0	10	5
1.25	13	5
1.5	15	5
2.5	20	5
4	25	5

Can I rewire my house myself?

Only you can answer this. There is no reason why you should not, provided you are reasonably capable and physically fit and that, if you live in rented accommodation, you have the landlord's permission.

However, you should appreciate that rewiring a house is quite a formidable task. Apart from the technical and safety angles, it involves considerable physical work such as lifting floorboards, climbing into roof spaces and, possibly, crawling under floorboards. Also, the work will probably be spread over a longer period than if done by professionals, particularly if you can only work in your spare time.

On the credit side you will certainly make a very substantial cost saving by doing the work yourself since the major part of a rewiring cost is for labour; you can also be confident that the work has been done properly.

You will need assistance for some of the work, particularly when installing cables under floors or in existing conduits.

If you decide that the work is beyond your ability, some electricity boards offer a reduction in the price if you do some of the preparatory work and making good yourself; if you wish to take advantage of this you should raise the question when they are preparing the quotation. You may be able to make a similar arrangement with an electrical contractor. Should you adopt this method it is preferable, but not essential, that your are available at all times whilst the rewiring is in progress.

How do I plan the rewiring?

You must first decide exactly what you intend to do, i.e. whether to rewire all items in their existing positions, or to add new lights or socket outlets or to move some items to new positions; also how much, if any, of the existing wiring can be retained.

It is a good idea to prepare a plan of the building and to show on it all the points which are to be rewired or added. You can then decide on the number of circuits, cable routes etc.

Can I use existing conduits?

Provided that they are in the right places you can use existing conduits when convenient, e.g. for the drops down to lighting switches and maybe to socket outlets. However, there is no particular advantage in using the conduits in roof spaces and below floors; the new cables may be run between or across joists, with mechanical protection where necessary.

When using existing conduits, the existing cables can be used to pull in the new cables; make the joint between the old and new cables so that the ends of all the wires are pointing to the *feeding* end; this reduces the possibility of the wires jamming in the conduit.

Existing conduits should *not* be used as the protective conductor; the bare wire in the PVC-sheathed cables should be used for this. When PVC-sheathed cables are installed in conduits, rubber or PVC bushes should be fitted at each end of the conduit and it does not matter if the conduits do not reach right up to the outlet points.

What programme should I use for rewiring?

This depends on personal choice and circumstances, but it is a good idea to start with the first-floor lighting. This is probably the easiest part, since the bulk of the work is in the roof space and you will gain experience and confidence.

Next, tackle all work under the floorboards of the first floor, i.e. ground-floor lighting, first-floor socket outlets, immersion heater and maybe the ground-floor sockets and cooker, depending on whether access is available for cabling below the ground floor. It is best to do all work in this area at the same time so as to avoid repeatedly lifting carpets and floorboards and moving furniture.

Finally, the work on the ground floor should be tackled, consisting mainly of the socket outlet ring circuit and cooker circuit.

When rewiring a particular floor or circuit, it is usually preferable to start at the end farthest from the supply point; this may reduce the temporary connections required at the end of each working period.

If you are moving to a fresh house it is a tremendous advantage if you can do the rewiring whilst it is unoccupied and unfurnished.

What system of wiring should I use?

As stated in Chapter 3 the most usual method of wiring domestic property is to use flat, PVC-insulated, PVC-sheathed cable with earth conductor (mainly twin and earth cable), and this system should be used whether or not the existing wiring is in conduit.

The loop-in system of wiring, with four-terminal ceiling roses (see Chapter 3) should preferably be used, but a joint box may be an advantage in odd cases to save long runs of cable.

Does rewiring involve being without electricity?

You should be without electricity only for short periods, possibly not at all. There are a number of ways of making the changeover to the new wiring, depending on circumstances.

If you already have a modern consumer unit which meets requirements, the new cables can be installed and connected to it in turn in lieu of the old wiring. It should not then be necessary for the electricity board to change their connections from the meter, provided the meter 'tails' meet their requirements, and there will be no interruption to the supply.

If the rewiring includes the installation of a new consumer unit you will need to call in the electricity board to disconnect the old switchgear and connect the new consumer unit. If space is available, or can be made available by moving the old switchgear temporarily, the best way is to fit the new consumer unit in its permanent position and for the electricity board to connect this, leaving the old switchgear also connected. The circuits can then easily be transferred to the new consumer unit as the rewiring proceeds and, on completion, the old switchgear can be disconnected by the electricity board and then removed. The above will involve two short interruptions to the supply.

If you are fitting a new consumer unit and the above scheme is not practicable, possibly because of lack of space, you will have to connect the new cables temporarily to the existing fuseboards etc.; then, when all the rewiring has been done, to arrange for the electricity board to disconnect your old switchgear. You can then remove this, fit the new consumer unit, and connect the cables to it; the electricity board can then make a further visit to connect the new consumer unit. The above changeover can easily be completed in a few hours and, provided the board make their first visit in the morning, the supply can be restored later the same day. The supply will, of course, be off between the two visits by the board. Advance arrangements should be made with the board.

What about temporary connections?

If the dwelling is occupied during the rewiring, you will probably need to make temporary connections at the end of each working period to provide essential services. You may also need temporary connections for such items as freezers, cookers etc. whilst they are being rewired.

It is essential that all temporary wiring and connections are left in a safe condition. Joints in wiring should be in proper screwed connectors, with no exposed live wires or terminals, and all earthing requirements must be met as for a permanent installation.

The above requirements apply in *all* cases, including any work in roof spaces or under floors; they also apply even if you make the connections last thing at night and intend to resume work next morning.

What materials shall I need?

The principal materials required for rewiring a normal dwelling are listed below. Other items should be added as necessary.

It is best to buy all, or the bulk of, the materials before you start the rewiring; this will avoid breaking off due to lack of essential items and will also help in shopping around for the best terms. British-made materials are preferable since, apart from other considerations, they are more likely to comply with the relevant requirements than imported items.

Cable is normally supplied in 50 m or 100 m coils. Don't buy too much initially – you can always get more if you need it. But for an average two-storey house you will probably need at least 100 m of 1 mm^2, flat, PVC-insulated, PVC-sheathed twin and earth cable for lighting and 50 m of 2.5 mm^2 similar cable for sockets etc. Larger cables will also probably be required for a cooker, shower etc. and the actual lengths should be ordered for these.

List of materials

Consumer unit (or units), fitted with fuses or MCBs.
1 mm^2 twin and earth flat PVC-sheathed cable for lighting.
1 mm^2 3-core and earth similar cable for lighting (alternatively use two twin and earth cables.)
2.5 mm^2 twin and earth similar cable for socket outlets etc.
4 mm^2, 6 mm^2 or 10 mm^2 twin and earth similar cable for cooker.
2.5 mm^2 and 6 mm^2 PVC-insulated single-core cable (green/yellow) for earthing and bonding.
Single-core PVC-insulated and sheathed cables for 'tails' from consumer unit to meter. Agree size with electricity board.
PVC sleeving, green/yellow and red.
Cable clips, plastic or buckle type.
Joint boxes, plastic 3- and 4-terminal.
Ceiling roses 4-terminal.
Lampholders, standard, white.
Lampholders, batten type.
Flex, circular, PVC-sheathed, heat-resisting, 2 or 3-core for luminaires.
Flex, 2.5 mm^2 similar to above for immersion heater.
Lighting switches, 1- or 2-way, rocker type, 5 A, for lighting.
Cord-operated ceiling switches, 1 or 2-way, for lighting and possibly for shower or immersion heater.
Outlet boxes, flush or surface, metal or insulated, 1- or 2-gang, depths as required. For lighting switches, socket outlets etc.
Socket outlets, 13 A, 1- or 2-gang, switched or unswitched, surface or flush mounting.
Fused connection units, surface or flush mounting.
Plugs, 13 A. Fused as required.
Switch, 20 A, DP, with pilot light, preferably engraved 'Water heater', or other type of switch as required.
Switch, 30 A or 45 A, DP, flush or surface mounting, for cooker; or

Cooker control unit, with 13 A socket outlet, flush or surface mounting (see Chapter 5).
Cable terminal box or outlet box, for concealed cable from control switch to cooker.
Screws, nails etc., as required.
Earthing clamps for connections to water or gas pipes etc.

Door bell, buzzer or chimes.
Mains transformer or batteries.
Bell push (or pushes).
Bell wire and staples.

You will also need the various tools required to carry out the work. And it's a good idea to lay some boards or planks across the joists in the roof space, particularly near the access trap. This could prevent you putting your foot through the ceiling. Temporary lighting in the roof space may also be well worthwhile.

What testing should I carry out?

As you complete each circuit, e.g. the first-floor lighting, you should immediately carry out the tests described in Chapter 10, particularly the polarity tests; this should preferably be done before you replace floorboards and carpets, in case defects are found.

The tests should also be repeated on completion of the rewiring, before calling in the electricity board.

What precautions should I take when doing the rewiring?

When doing work which involves crawling under the floor it is advisable that there should always be another person in the house in case of a possible accident. This also applies to a lesser extent to work in the roof space.

Some types of loft insulation, such as glass fibre, can cause irritation to the skin. When such insulation is present it is advisable to wear gloves, to keep your arms covered and also to wear a face mask. If irritation is experienced, it can be reduced by rinsing the affected areas under running water before washing.

Cables in roof spaces should not be in contact with, or be covered by, thermal insulation – see Chapter 3.

What points should I watch when lifting floorboards?

When lifting floorboards which have not been raised before, you will probably need to cut the tongue with a rip saw. Take care not to damage any cables or pipes beneath them.

Look out for boards which have been raised previously and use these if possible.

How can I locate a fault in the wiring?

Faults in wiring may be (a) a short circuit between phase and neutral or from phase or neutral to earth, (b) a disconnection (sometimes called open circuit) on any of the conductors, or (c) low insulation resistance between conductors or from a conductor to earth. The general principle in all cases is to localize the fault in steps until the actual cause is found.

However, before making any tests, examine the wiring as far as possible, particularly if any work has recently been carried out on the premises, e.g. screws or nails in walls may have penetrated hidden cables, or cables may have been damaged during other installation work.

As an example we will assume a phase-to-earth fault on a radial circuit feeding 13 A socket outlets. First, unplug all appliances from the circuit, switch off any fixed appliances, turn off the main switch on the consumer unit and remove the circuit fuse or MCB. Then connect the bell and battery set (see Chapter 10) between an earth connection and the phase terminal of the circuit in the consumer unit – see Fig. 51. The bell should then ring.

Next, remove a socket outlet near the centre of the circuit and separate the wires; then repeat the test at the consumer unit. If the bell still rings, the fault is between the consumer unit and the

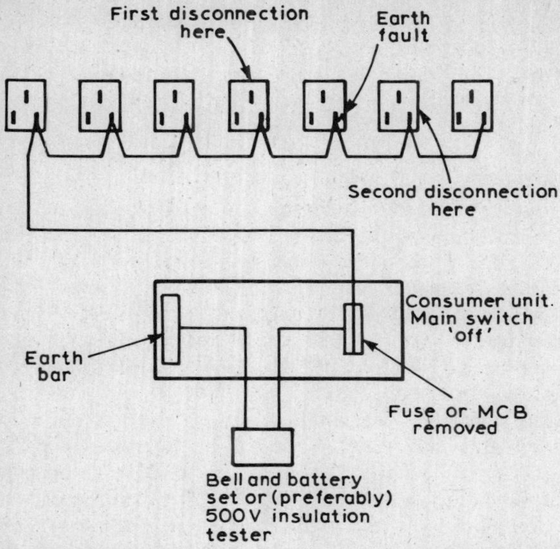

Fig. 51 Location of earth fault on radial circuit feeding socket outlets. Phase conductor only shown

socket outlet which is disconnected. If the bell does not ring, the fault is beyond the socket.

For the next test, remove a socket outlet near the centre of the faulty section and, from that point, make a test in each direction between phase and earth. The bell should ring for one direction but not the other, thus indicating the faulty section. This process should be repeated until the fault is located. It may be in a length of cable beween two socket outlets or (more likely) at an actual socket outlet or fused connection unit.

The above procedure can be applied to any type of radial circuit. For a ring circuit, the ring cables must be removed from their terminals in the consumer unit and separated from each other before making the tests.

It is possible that the above fault might not operate the bell and battery set due to the low test voltage (say 3–5 V), which might be insufficient to 'break down' the fault. In such a case, a 500 V insulation tester may be used to locate the fault, following the above procedure.

We will now assume a *disconnection* (or open circuit) on a radial circuit feeding 13 A socket outlets.

If the mains supply is connected, this can readily be located by plugging a test lamp into the socket outlets in turn, preferably starting at the supply end of the circuit and working in sequence along the circuit. A normal portable table lamp can be used. The lamp will, of course, light when plugged into sockets between the consumer unit and the fault, but not for sockets beyond the fault. Thus the fault can be localized and careful inspection will reveal the actual cause.

If the mains supply is not available, the tests may be made by means of a bell and battery set. Remove the circuit fuse or MCB and connect the bell and battery set to the phase and neutral terminals of the circuit at the consumer unit. Then at each socket outlet in turn connect the phase and neutral terminals together; this can be done easily by using a standard 13 A plug with the L and N terminals connected together. The fault can then be located as indicated above.

Disconnections on other radial circuits can be located using similar principles. For a ring circuit the cables must be disconnected and separated at the consumer unit before the tests are made.

Low-insulation faults cannot be located by using a bell and battery set, and a 500 V insulation tester must be used.

10

Testing

What tests are required?

The IEE Wiring Regulations detail the tests required on a new installation. Some of the tests require special and expensive testing equipment and also specialized knowledge, and are beyond the scope of most householders.

Before connecting a new installation or an extension to an existing system, the electricity board will normally carry out the necessary tests. The first test is free, but if defects are found and a further test is necessary, a charge may be made.

If the new wiring is fairly extensive it may be worthwhile, depending on your capability and confidence, to get an experienced electrician to check the installation and carry out the necessary tests before calling in the electricity board; he would, of course, require payment for this.

However, certain tests can be made quite easily by any normal person without expensive testing equipment, and these are detailed below.

Why must the continuity of a ring circuit be tested?

As indicated earlier (Chapter 5), 13 A socket outlets to BS 1363 are usually wired on the ring system with both ends of the ring terminating at the consumer unit. Since the ring is fed from both ends, a break in the circuit will not be apparent and the socket outlets will continue to work normally.

However, this situation must be avoided since the cable sizes for the ring are based on the assumption that the ring is fed from both ends and a break may well lead to overloading of the ring circuit cables, particularly if the break is near one end of the ring.

How is ring circuit continuity tested?

The continuity can easily be tested by means of a bell and battery set; this consists of a battery connected in series with a bell to two test leads (Fig. 52*a*). When the two test leads are connected together the bell rings. (A lamp may be used in lieu of a bell.)

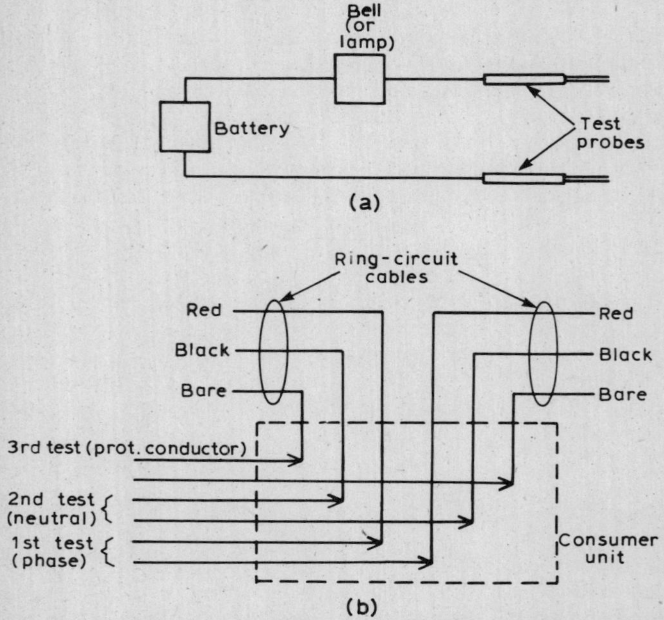

***Fig. 52** (a) Bell and battery set. (b) Ring circuit continuity test. Connections using bell and battery set*

The test should be made at the consumer unit with the supply disconnected. The two ring circuit cables are disconnected from the terminals and from each other and the tester is then connected in turn, to the two phase conductors, the two neutrals and the two protective conductors. In each case, the bell should ring; if it does not, a break in the ring is indicated. See Fig. 52*b*.

Why must polarity tests be made?

As explained in Chapter 2, it is essential that all SP switches and excess-current protective devices are connected in the phase conductor. It is also essential that the connections are correct at unswitched socket outlets and other outlet positions. If these conditions are not met, equipment may be 'live' even when the control switch is off or the fuse or MCB is removed.

How is a polarity test made?

Provided that the mains supply is available, tests at socket outlets can readily be made by using a standard 13 A plug with a mains voltage test lamp (filament type) connected between the phase (L) and earth terminals. The plug is inserted into each socket in turn, and glowing of the test lamp indicates that the phase and neutral connections are not reversed.

This test does not show up a reversal between the neutral and earth connections; this situation is not very likely but it must be avoided since, although the socket outlet will appear to work normally, there is a risk of overloading the protective conductor, which will be carrying the neutral current. You can check this by disconnecting the neutral conductors of the circuit under test at the consumer unit and repeating the above test at each socket. The test lamp should still glow; if not, a reversal of the neutral and protective conductors is indicated and this must be remedied. The neutral *must* be reconnected immediately after the test.

Polarity tests at positions other than socket outlets, such as lighting switches, Edison-screw lampholders and fused connection units, can be made by means of a mains-voltage test lamp connected via a lampholder to two test leads. A neon lamp is not suitable. In testing at lighting switches, remove the lamps, place switches to off and connect one side of the test lamp to the neutral or earth. The test lamp should glow when connected to one terminal of the switch but not the other. When testing at an Edison-screw lampholder remove the lamp but place the switch to on; the test lamp should glow only when connected to the *centre* contact of the lampholder.

Checks that MCBs or fuses are connected in the phase conductor can easily be made at the consumer unit, using the test lamp; and the connections of flexes to portable appliances can be checked by visual inspection or by simple tests between the plug and the appliance, using the bell and battery set.

How are polarity tests made if the mains supply is not connected?

In this case the test can be made by means of the bell and battery set described earlier, using one long and one short lead.

Connect the long lead to the main red phase cable at the consumer unit with the main switch on and all fuses or MCB's in position. Then take the tester and the short lead to each socket outlet in turn. Connect the test lead via a 13 A plug to the phase terminal and, if the socket is a switched type, the bell should ring when the switch is closed and stop when it is opened. At unswitched sockets, the bell should ring when the test lead is connected to the phase terminal.

To prove that there is not a neutral/earth reversal, connect the tester to the main neutral 'tail' on the supply side of the consumer unit and then connect the other side of the tester to the neutral terminal at each socket outlet. Ringing of the bell indicates that there is no neutral/earth reversal.

Tests at lighting points, switches etc. can be made in a similar manner as described earlier.

How can circuit continuity be tested?

Before the mains supply is connected it is advisable to check the continuity of circuits. This can be done very easily and quickly by connecting the bell and battery set to the main phase and neutral 'tails' at the consumer unit, with the main switch on and all fuses or MCBs in position. All lamps should be removed and fixed appliances switched off but all lighting switches and switches on socket outlets should be 'on'. The phase and neutral connections are then connected together in turn at each socket outlet, lighting point etc. and the bell should ring in each case.

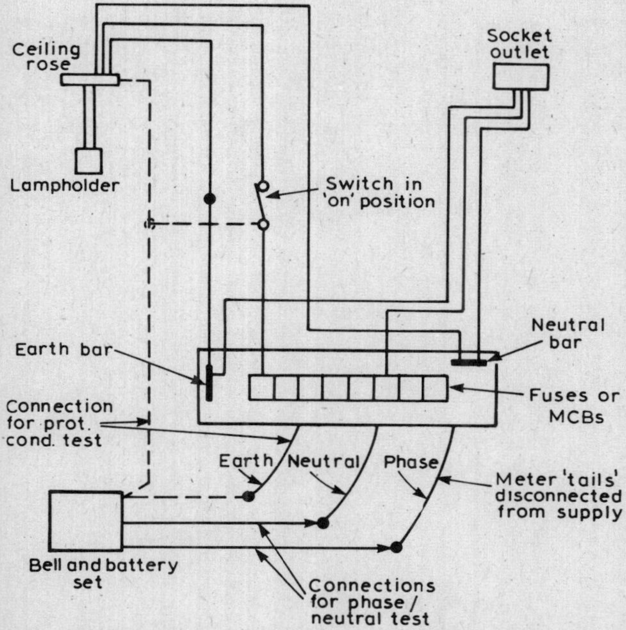

***Fig. 53** Connections for continuity test using bell and battery set*

The protective conductor is tested in a similar manner, the bell and battery set being connected to the protective conductor 'tail' at the consumer unit and the other side of the tester then connected in turn to the earth terminal or metalwork at *each* outlet point, including all switches, ceiling roses etc. Fig. 53 shows the connections for the above tests.

How is an earth-leakage circuit breaker (ELCB) tested?

The tests specified in the IEE Wiring Regulations require special equipment and knowledge and are beyond the scope of most householders.

However, as stated in Chapter 2, all ELCBs include a test push which simulates a fault and should cause the breaker to trip. This should be used to check the operation of the ELCB after installation and at regular intervals, say every month, afterwards. The breaker should trip instantly and isolate all loads supplied via its contacts.

Index